Socio-Affective Computing

Volume 11

Description

This exciting series publishes state-of-the-art research on socially intelligent, affective, and multimodal human-machine interaction and systems. It emphasizes the role of affect in social interactions and the humanistic side of affective computing by promoting publications at the crossroads between computer science, engineering and the human sciences (including biological, social, and cultural aspects of human life).

Three broad domains of social and affective computing will be covered by the book series: social computing; affective computing; and the interplay of these domains (for example, augmenting social interaction through affective computing).

Examples of the first domain include all types of social interactions that contribute to meaning, interest, and richness in our daily life, e.g., information produced by a group of people used to provide or enhance the functioning of a system. Examples of the second domain include computational and psychological models of emotions, bodily manifestations of affect (facial expressions, posture, behavior, physiology), and affective interfaces and applications, e.g., dialogue systems, games, and learning.

Research monographs, introductory- and advanced-level textbooks, and edited volumes are considered for the series.

Sancheng Peng • Lihong Cao

Textual Emotion Classification Using Deep Broad Learning

Springer

Sancheng Peng
Lab. Language Engineering and Computing
Guangdong University of Foreign Studies
Guangzhou, China

Lihong Cao
School of English Education
Guangdong University of Foreign Studies
Guangzhou, China

ISSN 2509-5706 ISSN 2509-5714 (electronic)
Socio-Affective Computing
ISBN 978-3-031-67717-5 ISBN 978-3-031-67718-2 (eBook)
https://doi.org/10.1007/978-3-031-67718-2

This Springer imprint is published by the registered company Springer Nature Switzerland AG
The registered company address is: Gewerbestrasse 11, 6330 Cham, Switzerland

Preface

In the recent decade, we have witnessed the dramatic development of natural language processing (NLP in short), not only for relevant technologies of NLP but also for various practical applications of human society. The rapid development of deep learning has pushed NLP into a new spring after a long winter. Nowadays, there are numerous successful examples of significant progress in various domains with the help of NLP, for example, the application of NLP in sentiment analysis, machine translation, machine Q&A, smart justice, automated essay scoring, and so on.

As a flat network, there are many advantages in broad learning, such as simple network structure, short training time, and good generalization performance, which has become an alternative solution for the deep learning models. At present, some researchers have utilized broad learning for many applications, such as nonlinear system identification, target location, image processing, automatic control, fault diagnosis, and emotion classification.

We classify the textual emotion classification research using deep broad learning into three categories: single domain textual emotion classification based on general and variant broad learning, cross-domain textual emotion classification based on deep broad learning, and emotion classification in textual conversations based on deep broad learning.

Thus, this book is organized into nine chapters with four parts. The first part introduces the general and variant broad learning for the textual emotion classification task. In this part, Chap. 2 introduces the general broad learning, and modeling process of emotion classification based on the bidirectional encoder representation from transformers (BERT) and broad learning. Chap. 3 depicts cascading broad learning, which combines the permuted pre-trained language model (MPNet) for the textual emotion classification. Dual broad learning is adopted to conduct the textual emotion classification in Chap. 4.

The second part focuses on single-source domain adaptation for emotion classification modeling by combining deep learning and broad learning. We introduce single source cross-domain emotion classification on the basis of broad learning and convolutional neural networks in Chap. 5. We further introduce multi-source domain adaptation for emotion classification by using broad learning and bi-directional long short-term memory networks in Chap. 6.

The third part introduces emotion classification in textual conversations by combining deep learning and broad learning. Based on broad learning, convolutional neural networks, and bi-directional long short-term memory networks, a new method is presented to capture the local contextual information and global contextual information in Chap. 7. In addition, based on relational graph attention network and broad learning, another approach is proposed to the contextual features and long-distance dependencies in Chap. 8.

The last part focuses on the future advancement of emotion classification research. On the basis of the above description in Chaps. 1–8, we discuss the challenges and future research trends in Chap. 9. Each chapter of the book is relatively independent, and the readers would select any order to read or browse the book according to their demands.

Emotion classification is an important research field, and adopting broad learning or deep broad learning for emotion classification is a newer sub-field. This book aims to introduce interested readers to this promising research field, and to design an introductory subject for senior undergraduate and postgraduate students, researchers, and industrial engineers, as well as anyone interested in natural language processing, especially emotion computing. The readers are expected to have basic knowledge in machine learning, pre-training model, and deep learning. Broad learning could be good prior to read this book, but not mandatory.

We expect the book to bring the readers an overview of emotion classification, and pave a starting ground for them to further explore the uncharted land of the domain in both theoretical and application perspectives. Because of the limitation of time and knowledge, we think there are many possible problems and errors. We humbly invite you to point them out to us, so as to correct them in our future versions.

We sincerely hope you will enjoy reading the book, and let us work together to explore the promising uncharted land.

Guangzhou, China
January 2024

Sancheng Peng
Lihong Cao

Acknowledgments

This book is a product of our research process on emotion computing, especially in the rapid development of deep learning and broad learning. We sincerely appreciate the suggestions, help, comments, and guidance of our colleagues and friends during the writing. Since there are so many of them, it would be a long list if we list their names here. We want to say "Thank you."

We would like to thank the members of our research groups at the Guangdong University of Foreign Studies. Master's students and undergraduate students are key players in our research, and they helped establish a solid foundation for the content of the book through numerous meetings and discussions. Some of them contributed significantly in the drafts: Rong Zeng for Chaps. 2, 4, and 8, Guanghao Chen for Chap 8, Xiangyu Nie and Changli Deng for Chaps. 5 and 6. In addition, we would like to thank Prof. Chengqing Zong helping to proofread Chaps. 5 and 6, and Prof. Shui Yu helping to proofread Chaps. 1 and 3.

We started the plan for this book in 2021 and finished the preparation in 2023. We sincerely thank the professional staff at Springer, especially Dr. Alexandru Ciolan and Sneha Arunagiri, for providing insightful suggestions and professional support during the preparation of this book; and Dr. Nick Zhu for offering invaluable help during the evaluation of the book proposal.

We are also grateful to our organizations, Laboratory of Language Engineering and Computing at Guangdong University of Foreign Studies, which have offered outstanding support and facilities to prepare this book.

Last but not least, we would like to thank our families, for their continuous support, encouragement, and understanding.

Contents

Acronyms

AAAI	Association for the Advancement of Artificial Intelligence
ACM	Association for Computing Machinery
ACAN	Adversarial category alignment network
BERT	Bidirectional encoder representation from transformers
BBL	Bidirectional encoder representation from transformers and broad learning
Bi-LSTM	Bidirectional long short-term memory
BL	Broad learning
CBL	Cascading broad learning
CDSC	Cross-domain sentiment classification
CNN	Convolutional neural network
CNNBL	CNN-based broad learning
CTDA	Contrastive transformer-based domain adaptation
DAE	Denoising autoencoder
DAML	Deep adversarial mutual learning
DAN	Deep averaging network
DBL	Deep broad learning
DL	Deep learning
DNN	Deep neural network
	Diagram graph network
DialogueGCN	Dialogue graph convolutional network
DAG	Directed acyclic graph
DIC	Domain-invariant classifier
DIF	Domain-invariant feature
DSC	Domain-specific classifier
DSF	Domain-specific feature
DANs	Dual attention networks
DAS	Dual attention suppression
DaViT	Dual attention vision transformers
DualCNN	Dual convolutional neural network
DualGCN	Dual graph convolutional networks

DDGNet	Dynamic dual graph networks
ECTC	Emotion classification in textual conversations
EL	Emotion lexicon
EmNet	Emotional network
FN	False negative
FP	False positive
GRNN	Gated recurrent neural networks
GRU	Gated recurrent unit
GANs	Generative adversarial nets
GAT	Graph attention network
GCN	Graph convolutional network
HATN	Hierarchical attention transfer network
HiGRU	Hierarchical gated recurrent unit
IEEE	Institute of Electrical and Electronics Engineers
IRER-EA	Interpretable relevant emotion ranking with event-driven attention
JBNN	Joint binary neural network
KAITML	Knowledge aware incremental transformer with multi-task learning
KET	Knowledge-enriched transformer
KL	Kullback Leibler
LM	Language model
LSTM	Long short-term memory
ML	Machine learning
MPNet	Masked and permuted language modeling
MLM	Masked language model
MOOCs	Massive open online courses
MMD	Maximum mean discrepancy
MoE	Mixture-of-experts
MF-DAKT	Multi-factors aware dual-attentional knowledge tracing
MHA	Multi-head attention
MSAN	Multi-source adaptation network
MBL	Multi-source BL
MDA	Multi-source domain adaptation
MVN	Multi-view network
NB	Naive Bayes
NLP	Natural language processing
NSP	Next sense prediction
PLM	Permutation language model
PBLM	Pivot-based language model
Q&A	Question & answer
RBF	Radial basis function
RNN	Recurrent neural network
RGAT	Relational graph attention network
RKHS	Reproducing kernel Hilbert space
SDA	Scaled dot product attention

SAL	Selective adversarial learning
SCP	Significant consistent polarity
SSDA	Single-source domain adaptation
SATL	Softly associative transfer learning
SINN	Speaker influence aware neural network
SVM	Support vector machine
TEC	Textual emotion classification
TGN	Textual graph network
TODKAT	Topic-driven and knowledge-aware transformer
TN	True negative
TP	True positive
UMAN	Universal multi-source adaptation network
UDA	Unsupervised domain adaptation
WTN	Wasserstein-based transfer network
WDIRL	Weighted version of the domain invariant representation learning

Chapter 1
Introduction

Abstract Textual emotion classification (TEC) aims to extract and analyze user emotional states in texts. There has been a rapid development of various deep learning (DL) methods that have proven successful in many domains such as audio, image, and natural language processing. This trend has drawn increasing numbers of researchers from traditional machine learning to DL for their scientific research. In this chapter, we provide an overview on TEC based on DL methods. After introducing a background of TEC that includes defining emotion, emotion classification methods, and application domains of TEC, we provide an overview on the DL technology.

1.1 Textual Emotion Classification

1.1.1 Overview

With the rapid development of communication technology, various social network [56, 63] platforms have become an essential medium for people. They can use mobile phones, computers and other terminal devices to share their information on various network platforms, and further to express their opinions anytime and anywhere, such as product reviews [64], news reviews, and movie reviews. Thus, a large account of texts have emerged on Internet, and most of them contain emotional information.

TEC can help to understand the views and positions of users, and also can help enterprises, research institutions, and decision-making department of governments to grasp emotion trend of society [11]. Emotion classification [36, 44] focuses on automatically extracting the emotional states (i.e., joy, love, anger, sadness, fear, evil, and desire) of users from their reviews [16, 39, 49] (e.g., blogs, reviews, tweets, etc.) in social networks.

TEC is the task of extracting and analyzing user emotional states in texts. TEC not only acts as a standalone tool for information extraction but also plays an important role for various natural language processing (NLP) applications, including e-commerce [7], public opinion analysis [22], big search [19], information

S. Peng, L. Cao, *Textual Emotion Classification Using Deep Broad Learning*,
Socio-Affective Computing 11, https://doi.org/10.1007/978-3-031-67718-2_1

prediction [38], personalized recommendation [61], healthcare [15], and online teaching [50].

The idiom of "seven emotions and six desires" refers to joy, love, anger, sadness, fear, evil, and desire. Among them, only a few are positive emotions and the rest are negative, indicating that people are naturally more sensitive to negative emotions. In real-world life, negative emotions are also easier to propagate than positive emotions. The 2011 Annual Report of China's Internet Public Opinion Index [2, 66] stated "In 2011, there are more than 80% negative events for the total number of topics. The negative events on Microblog and Tianya forum accounted for 75.6% and 95.8%, respectively, which are higher than those of other social media."

With the rapid development of social networks [9, 39, 40, 63], people have changed from general users to network information producers. According to the 38th statistical report on the development of China's Internet published by China Internet Information Center [3], the number of Internet users in China has reached 710 million, and the Internet penetration rate reached 51.7% in June 2016. Among them, 656 million were mobile Internet users, 242 million used Microblog, and more than 100 million had daily blogs. Among this massive number of short text messaging, negative emotions are the most prevalent.

Early research for TEC task focused on either a positive/negative bipartition or a positive/negative/neutral tripartition of emotion analysis [8, 31]. However, such partitioning ignores subtle user emotion changes and their psychological states, preventing a full expression of people's complex inner emotional world. This gave rise to bipartition-oriented emotion analysis named "sentiment analysis" [10, 51], while more encompassing emotion analysis was dubbed as "fine-grained sentiment analysis".

Existing TEC method may be divided as follows [42]: text-oriented monolingual, text-oriented cross-linguistic, and emoji-oriented cross-linguistic. Text-oriented monolingual method [41] focuses on classifying textual emotion in a single language (e.g., Chinese). Text-oriented cross-linguistic method [4] focuses on exploring TEC from source language text (e.g., English) to target language (e.g., German). Emoji-oriented multi-linguistic method [6] focuses on predicting emoji in multi-linguistic texts (e.g., English and Spanish).

In 2012, DL models [47, 65] were introduced to NLP after they achieved successful object recognition via ImageNet [17]. DL methods improved on statistical learning results in many fields. At present, a neural network-based NLP framework has achieved new levels of quality and become the dominating technology for NLP tasks, such as sentiment analysis, machine translation, and question answering systems.

Popular DL methods are utilized to TEC task, including deep averaging network (DAN) [28], denoising autoencoder (DAE) [54], convolutional neural network (CNN) [30], recurrent neural network (RNN) [20], long short-term memory (LSTM) network [24], bidirectional long short-term memory (Bi-LSTM) network [48], gated recurrent unit (GRU) [14], bidirectional gated recurrent unit (Bi-GRU) [55], attention [27], multi-head attention (MHA) [35, 52], generative adversarial network (GAN) [21], graph convolutional network (GCN) [23, 29], and graph attention network (GAT) [53]. In general, researchers may integrate multiple methods into a specific framework (rather than just one) to improve TEC performance.

1.1.2 Emotion Overview

Emotion Definition

In this subsection, we will explain the meaning of emotion. The basic understanding of emotion is summarized as follows.

Definition 1.1 Emotion [1] is defined by the Merriam-Webster Dictionary as "A conscious mental reaction (e.g., anger or fear) subjectively experienced as strong feeling usually directed toward a specific object and typically accompanied by physiological and behavioral changes in the body".

Definition 1.2 Emotion [25] denotes people's attitude experience and corresponding behavioral responses to objective things.

Definition 1.3 Emotions [26] are "generated states in humans that reflect evaluative judgments of the environment, the self and other social agents".

Definition 1.4 Emotion [36] is the feeling or reaction that people have due to a certain event. "Happy", "sad", "angry", "fear" are few examples of emotions that can be expressed. "Emotions" and "sentiments" are often considered interchangeable, yet the latter represents emotional polarity or general emotional states (i.e., positive, negative, or neutral).

Definition 1.5 Emotion [34] is a mental state that arises spontaneously rather than through a conscious effort; it is also often accompanied by physiological changes.

Definition 1.6 Emotion [45] is often defined as an individual's mental state associated with thoughts, feelings, and behavior.

Emotion Type

In this subsection, we will introduce the classification of emotion. The basic understanding of emotion types is surveyed as follows.

Ekman [18] categorized emotion into six basic types: anger, disgust, fear, happiness, sadness, and surprise.

Parrott [37] adopted a tree structure to describe emotion with six kinds: love, joy, surprise, anger, sadness, and fear.

Plutchik [43] visualized a wheel-shaped emotion classifier (based on the Ekman's model [18] with two additional categories) with four bipolar sets: joy and sadness; anger and fear; trust and disgust; and surprise and anticipation.

Lin Chuanding [33], a Chinese modern psychologist, divided emotion into 18 Shuowen-based categories: joy, quiet, caress, worry, fright, pity, fear, grief, shame, sorrow, anger, vexation, reverence, hatred, arrogance, greed, jealousy, and shame.

These existing emotional category approaches focus on modeling emotions based on distinct emotion classes or labels. These models assume that the discrete emotion categories exist in texts.

1.1.3 TEC Application

TEC has been widely studied in opinion mining, psychology, neuroscience, and behavioral science, as emotion is an important element of human nature. Such analysis plays an important role in many application fields like e-commerce, public opinion analysis, big search, information prediction (e.g., financial prediction, presidential election prediction), personalized recommendation, healthcare (e.g., depression screening), and online teaching. The specific framework is shown in Fig. 1.1.

E-commerce

With the rapid development of mobile Internet, online shopping has become more popular. Users often provide their personal comments on products purchased via Taobao, Jingdong, Amazon, and other e-commerce platforms. Bharti et al. use the product reviews [7] to conduct real-time TEC, which can help to obtain useful emotional and behavioral consumer characteristics, and to predict the trend changes in consumer preferences. Such information would help a majority of consumers deeply understand the quality of goods, pre-sale and after-sales services, logistic services, and other relevant information, guiding them through their future purchases. Manufacturers would also benefit from first-hand consumer feedback, timely product shortage warnings, and improved product quality and design. Sellers would benefit from knowing consumer psychological states as they relate to

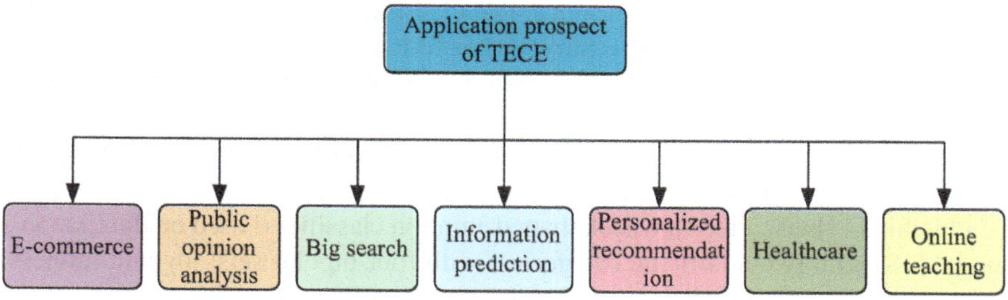

Fig. 1.1 The application prospect of TECE

available commodities and related services. Sellers who can capture consumer psychology can make timely sales, purchase, and marketing decisions, allowing them to reach market dominance.

Public Opinion Analysis

Network public opinion [22] refers to different views of popular social issues expressed on the Internet. This public form of social opinion carries appreciable influence and allows tendentious opinions of issues affecting reality via the Internet. Public opinion analysis is utilized to objectively reflect the state of public opinion by collecting and sorting out people's attitudes as well as discovering relevant opinion tendencies. Many irrational emotions (such as negative feelings towards the rich, official, powerful, or corruption) are expressed and strengthened by means of Internet violence or entertainment, driving people towards more extreme emotional reaction. Irrational netizen emotion causes national and societal security risks. Thus, relevant national management departments require knowledge of network public opinion trends to guide that opinion properly and in a timely fashion. However, when such information is obtained through various channels, its complexity prevents manual processing. This shortfall makes developing accurate and effective emotion analysis systems significant as well as the automatic processing of network public opinion information necessary to maintain the national security and social stability.

Big Search

With the network space expansion, network application mode development and the arrival of the big data era, the Internet has become ubiquitous and has given rise to big search technology [19]. Big search is becoming a strong tool and catalyst for the network development. Big search, the next generation search engine for cyberspace, is becoming an urgent need. Compared with traditional search, big search can understand user search intentions on a semantic level while also perceiving user needs according to their spatio-temporal location, emotional state, and historical preferences. Big search can also remove the false data and protect the user privacy. In addition, big search can provide an intelligent solution to users, making retrieval technology based on TEC an important research task.

Information Prediction

As Internet has developed, larger numbers of people rely on it for information and communication sharing, particularly for social network interactions (e.g., Microblog, Wechat, stock, and futures forums). Emotion analysis technology can be used to analyze the impact of social networks upon user lives and predict developing trends by way of the commentary, news articles, and other content. The main applications of information prediction are listed as follows:

1. Financial prediction
   More and more financial investors are turning to network for their financial information obtaining and investment opinion exchanging. This makes professional forums on subjects such as stocks and futures rich with financial data and investor emotional information (i.e., important factors affecting investor behavior and psychology). Behavioral finance dictates that the psychology and behavior of irrational stock investors affect the stock market situations, causing stock prices to deviate from their correct value. Thus, the stock market volatility can be predicted by reviewing the investors' emotion and behavior information drawn from a stock forum.
2. Election prediction
   TEC plays an increasingly important role in the prediction of democratic elections. Paul et al. [38] designed a framework, called Compass, adopting the 2016 U.S. presidential election as an example to analyze the election-related crowd emotion. They built a spatial-temporal emotional map through Compass for the election, and applied the map to match the election results to an extent. The study showed that any political event can be described by its popularity in negative and positive senses. In addition, Ceron et al. [12] utilized TEC to calculate the support rates for the political leadership candidates in the 2011 Italian parliamentary election and the 2012 French presidential election.
3. Other prediction
   TEC can also be applied to predict public opinion regarding various policy events (such as personal income tax adjustment, medical insurance reform, and retirement delays) as well as provide support for the national policy formulation. In addition, TEC can be applied to predict and judge natural disasters like typhoon, epidemics [5] and earthquakes. With the application of information prediction, TEC has attracted more and more attention, which helps to predict the developing trend of events by analyzing the Internet news, blogs, and other information sources.

Personalized Recommendation

The personalized recommendation systems [13, 60, 62] can provide users with a tool to address information overload issues. However, the traditional recommendation technology only considers overall user scores while ignores the emotional information contained in user comments. Such commentary usually contains subjective user views, preferences, and emotions regarding certain attributes of things, reflecting user emotional tendencies for those attributes. Mining and exploiting user commentary to its fullest extent gives rise to more accurate personalized recommendation while helping address the following issues like cold starts, data sparsity, and low recommendation accuracy.

Healthcare

According to the text recorded from social networks on psychological counseling, the emotional state information of patients suspected to be depressed was analyzed and screened using TEC [15]. Yang et al. [59] utilized the disease analysis interview corpus from the University of Southern California to analyze whether a patient with poor mental health (e.g., those diagnosed with anxiety or PTSD) also suffered from depression. To lighten the psychological burden of interviewees, interviews were conducted by a human-controlled robot. To effectively determine possible depression conditions, it needs to collect data like the recorded texts and PHQ-8 questionnaires.

Online Teaching

With the popular application of massive open online courses (MOOCs) [32, 58], a large number of online courses and reviews have been generated. Most reviews contained the emotions and opinions on the online courses from students. Tucker et al. [50] found a positive correlation between student emotional tendencies from their forum-based reviews and their learning performance on the MOOC platform. Thus, using TEC to analyze comment information on the MOOC platform for obtaining the course-related emotion information can help teachers find problems in curriculum arrangement, knowledge systems and teaching methods, enabling timely teaching plan and method optimization to further improve the teaching quality and students' learning efficiency.

1.2 Deep Learning

In this subsection, we will provide an overview of relevant DL models. The basic understanding of relevant models are summarized as follows.

1.2.1 DL Overview

DL model focuses on the feature learning. In essence, it aims to learn the complex feature representation on the basis of the original feature input, through multi-layer nonlinear processing. If combined with specific domain tasks, DL can construct new classifiers or generate tools through the feature representation of automatic learning, and realize the domain-oriented classification or other tasks. The specific description of the algorithm for a DL model are listed as follows [46, 57]:

Step 1: Construct a learning network through random initialization, set the total number of network training layers n, initialize unlabeled data as the input set of network training, and initialize the training network layer $i = 1$.

Step 2: Based on the input set, an unsupervised learning algorithm is adopted to pre-train the learning network of the current layer.

Step 3: The training results of each layer are output as the input for the next layer; construct the input set once again.

Step 4: If i is less than n, then $i = i + 1$, and return to Step 2; otherwise, proceed Step 5.

Step 5: The supervised learning algorithm is applied to adjust the network parameters of all layers, forcing any errors to satisfy practical requirements.

Step 6: Complete the classifier construction (such as neural network classifiers) or complete the construction of the deep generation model (such as a deep neural network (DNN)).

1.2.2 Deep Averaging Networks (DAN)

DANs are constructed by stacking the nonlinear layers over the traditional neural bag-of-words models. For each document, a DAN takes the arithmetic mean of the word vectors as the input, and passes it through one or more feed-forward layers until there is a softmax for classification. The specific framework of a DAN is shown in Fig. 1.2.

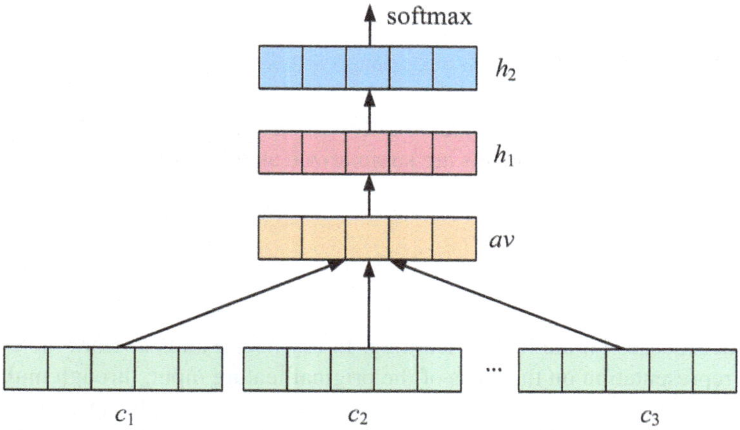

Fig. 1.2 A DAN with two feed-forward layers

According to the text classification, a DAN needs to map an input sequence of n tokens to one of k labels, and it also needs the following three steps:

Step 1: Take the vector average of the embedding associated with the input sequences of tokens n:

$$av = \sum_{i=1}^{n} c_i / n, \quad (1.1)$$

where c_i denotes the word embedding sequence.

Step 2: Pass the average through one or more feed-forward layers. If there is one layer, the average word embedding is represented as follows:

$$h_1 = f(w_1 \cdot av + b_1), \quad (1.2)$$

where b denotes the offset term and w denotes the $k \times d$ weight matrix. If there are more layers, the average word embedding is represented as follows:

$$h_i = f(w_i \cdot h_{i-1} + b_i). \quad (1.3)$$

Step 3: Conduct (linear) classification on the representation of the final layer:

$$\text{soft}\max(q) = \frac{\exp q}{\sum_{j=1}^{k} \exp q_j}. \quad (1.4)$$

1.2.3 Denoising Autoencoder (DAE)

A DAE is an unsupervised learning algorithm that acts as an autoencoder modification. It can form a DL network with multiple stacked layers. A DAE consists of three layers like encoder, hide, and decoder. The specific framework of a DAE is shown in Fig. 1.3.

Encoder $f(\bar{x})$ aims to reduce the dimensionality of the high-dimensional input. Input x is added noise to obtain a destroy version $\bar{x}$, which is input into $f(\bar{x})$. Implicit coding result y is then obtained through the linear transformation and activation function. Decoder $g(y)$ aims to obtain a reconstructed vector z. The specific computing for y and z is represented as follows:

$$y = f(\bar{x}) = S_f(W\bar{x} + b_y), \quad (1.5)$$

$$z = g(y) = S_g(W^T y + b_z), \quad (1.6)$$

$$S_f = \text{sigmoid}(x) = 1/(1 + e^{-x}), \quad (1.7)$$

$$S_g = \text{sigmod}(y) = 1/(1 + e^{-y}), \quad (1.8)$$

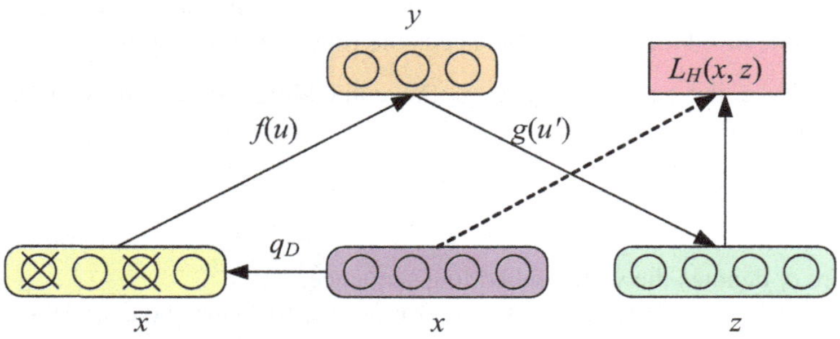

Fig. 1.3 The framework of DAE

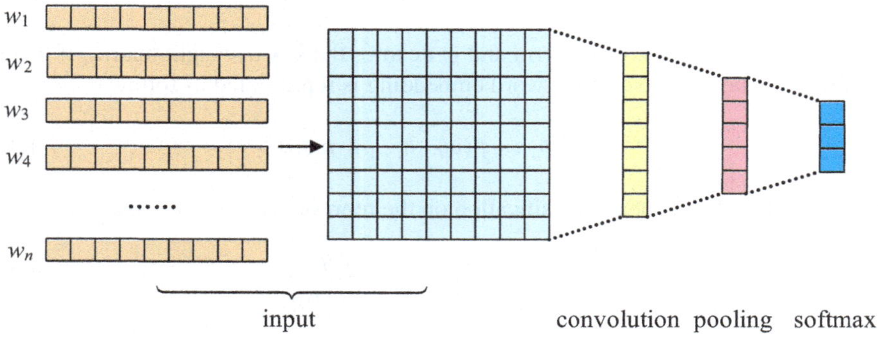

Fig. 1.4 The framework of CNN

where S_f denotes an activation function of the linear transformation, S_g denotes an activation function of the decoder, b denotes the offset term, and W denotes the weight matrix.

1.2.4 Convolutional Neural Network (CNN)

A CNN is a kind of feed-forward neural network, mainly including five layers like input, convolution, pooling, full connection, and output. The specific framework of a CNN is shown in Fig. 1.4.

In NLP, the application steps of a CNN mainly include: taking a vectorized sentence matrix as the input, convoluting the original input with multiple convolution kernels in the convolution layer, and obtaining multiple feature representations of the original input. Extracted features are then sent through the pooling layer as input and sampled to obtain more abstract features. Finally, through the full connection layer, the corresponding classification function is adopted to classify results to complete the classification task.

Input Layer It is responsible for the vectorization of input data. For a length n of a given sentence, the matrix of input layer can be represented as:

$$E \in \mathbb{R}^{n \times m}, \tag{1.9}$$

where m denotes the dimensionality of a word vector.

Convolution Layer It uses different convolution kernels to perform the convolution operation for the input matrix, extract local features from the input, and obtain feature maps of the convolution kernel. The specific representation is represented as follows:

$$X^l_j = f(\sum\nolimits_{i=1}^{M} X_i^{l-1} * K^l_{ij} + b^l_j), \tag{1.10}$$

where X^l_j denotes the output of convolution operation between the input and the j-th convolution kernel, X_i^{l-1} denotes the output of the previous layer (i.e., the i-th local receiving domain of the current convolution layer l), K^l_{ij} denotes the j-th convolution kernel of l, b^l_j denotes the offset term of the current convolution kernel, and M denotes all locally received domains that the convolution kernel needs to traverse.

Pooling Layer It uses the corresponding sampling function to sample the characteristics of the matrix generated by the convolution operation, and extracts more important features to reduce the matrix dimensionality. In addition, it simplifies the calculation process while preventing key features from being discarded. In general, the common pooling algorithm includes max pooling and average pooling.

Full Connection Layer It classifies the input, obtains the classification results, and is responsible for taking those results to the output layer.

1.2.5 Recurrent Neural Network (RNN)

A RNN is a kind of feed-forward neural network with a ring structure and a specific memory function. Its input includes current input samples as well as information obtained in the previous time, so that information can be cycled in the network at any time. The specific framework of a RNN is shown in Fig. 1.5, in which x denotes the input layer, O denotes the output layer, H denotes the hidden layer, and u, V, and w denote the weights of the above each respective layer. The output of the hidden layer is represented by:

$$h_t = \sigma(W_{xh}x_t + W_{hh}h_{t-1} + b_h). \tag{1.11}$$

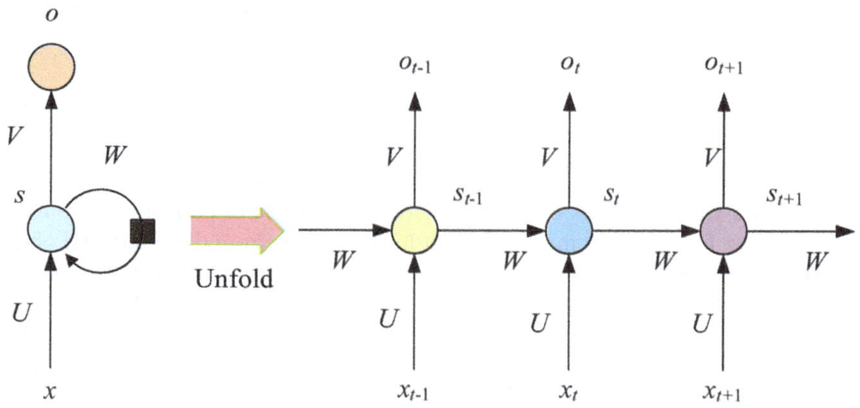

Fig. 1.5 The framework of RNN

The output of the out layer is represented by:

$$o_{t+1} = \sigma(W_{hy}h_t + b_y), \tag{1.12}$$

$$y_t = \text{softmax}(o_t), \tag{1.13}$$

where y_t denotes the output in time t, W_{hh}, W_{xh}, and W_{hy} denote the weight matrix, b_h and b_y denote the bias term, h_t denotes the hidden state in t, and x_t denotes the input data.

1.2.6 Long Short-Term Memory (LSTM)

An LSTM network is a kind of RNN that can learn the relationship of long-term dependency, characterize the information of a time sequence, and effectively solve the problem of gradient vanishing or gradient exploding faced in RNN training. It was first proposed by Hochreiter and Schmidhuber in 1997 [24]. After that, many researchers have optimized and improved upon it, leading to rapid development and its wide applications in various NLP fields.

Each unit of an LSTM network consists of four components like a memory cell, input gate, output gate, and forget gate. Memory cells are connected circularly with each other. Three nonlinear gate cells can be utilized to adjust the information of memory cell input and output flows. The specific framework of an LSTM network is shown in Fig. 1.6.

The forward computing process of a LSTM network is represented as follows.

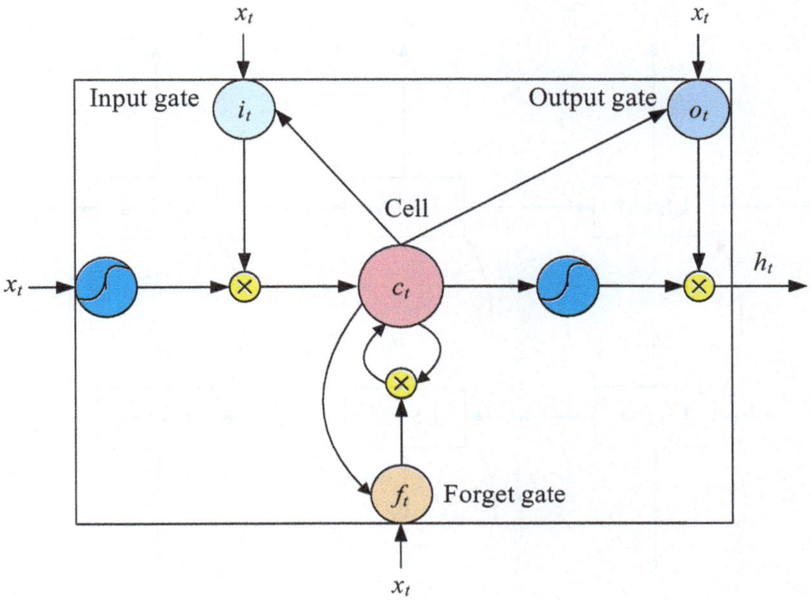

Fig. 1.6 The framework of LSTM

Input gate:

$$i_t = \sigma(W_i x_t + U_i h_{t-1} + b_i), \tag{1.14}$$

Forget gate:

$$f_t = \sigma(W_f x_t + U_f h_{t-1} + b_f), \tag{1.15}$$

Memory cell:

$$c_t = f_t \bullet c_{t-1} + i_t \bullet \tanh(W_c x_t + U_c h_{t-1} + b_c), \tag{1.16}$$

Output gate:

$$o_t = \sigma(W_o x_t + U_o h_{t-1} + b_o), \tag{1.17}$$

Result output:

$$h_t = o_t \bullet \tanh(c_t), \tag{1.18}$$

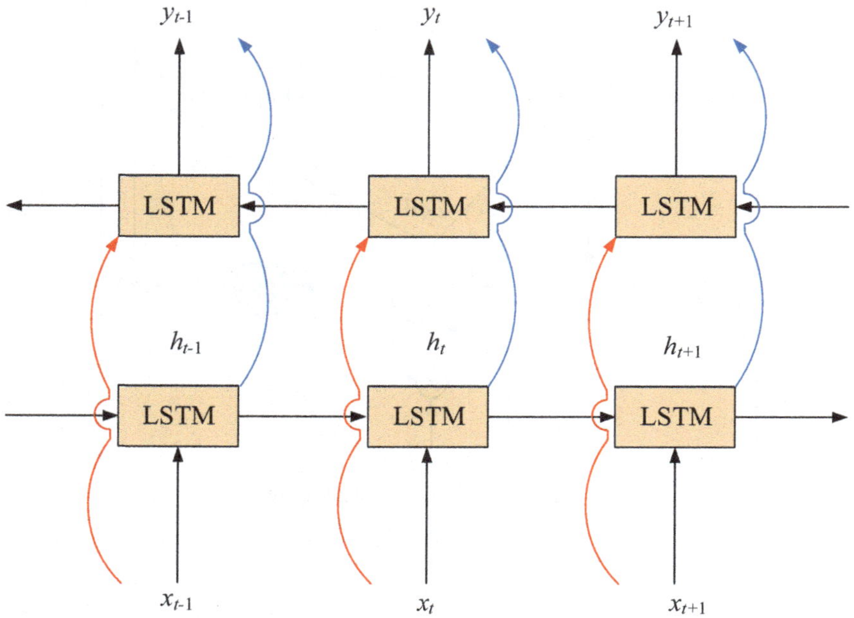

Fig. 1.7 The framework of Bi-LSTM

where x_t denotes the input vector (such as a word vector) at time t; f, i, and o denote the activation vectors of the forget gate, input gate, and output gate, respectively; c denotes the memory unit vector, h denotes the output vector the LSTM unit, $W_i, U_i, W_f, U_f, W_c, U_c, W_o$, and U_o denote the weight matrix; b_i, b_f, b_c, and b_o denote the bias vector; and σ and tanh denote the activation function.

1.2.7 Bi-directional Long Short-Term Memory (Bi-LSTM)

A Bi-LSTM is an improved LSTM model. One-directional LSTM uses the previous information to deduce the subsequent information, requiring information processing and preventing it from accessing future context or integrating context information, which affects system prediction performance. A Bi-LSTM uses two LSTM networks to train together and start their respective sequences from the opposite ends while being connected back to the same output layer. Thus, it can integrate the past and future information of each point. A Bi-LSTM includes the forward and backward calculation. The horizontal direction represents the bi-directional flow of the time sequence, and the vertical direction represents one-directional flow from the input layer to the hidden layer and on to the output layer. The specific framework of a Bi-LSTM is shown in Fig. 1.7.

The forward calculation of hidden vector $\boldsymbol{h}$ is represented as follows:

$$\boldsymbol{h}_t = \text{LSTM}(x_t, \boldsymbol{h}_{t-1}). \tag{1.19}$$

The backward calculation of the hidden vector $\overleftarrow{h}$ is represented as follows:

$$\overleftarrow{h}_t = \text{LSTM}(x_t, \overleftarrow{h}_{t-1}). \tag{1.20}$$

The output is represented as follows:

$$y_t = g(W_{\boldsymbol{h}y}\boldsymbol{h}_t + W_{\overleftarrow{h}y}\overleftarrow{h}_t + b_y), \tag{1.21}$$

where x_t denotes the input data, y_t denotes the output at time t, $W_{\boldsymbol{h}y}$ and $W_{\overleftarrow{h}y}$ denote the weight matrix, and b_y denotes the bias term.

1.2.8 Gated Recurrent Unit (GRU)

A GRU is an LSTM variant. Since LSTM can overcome the problem of gradient vanishing or gradient exploding when dealing with the relationship of long-distance dependency, and it can keep the dependency relationship of long-distance and short-distance for temporal data. A GRU retains these advantages while offering a simpler network structure. In comparison with a LSTM network's three-gate structure, there are only two gates (i.e., an update gate and a reset gate) in a GRU. The specific framework of a GRU is shown in Fig. 1.8.

The forward calculation of a GRU is represented as follows:

Update gate:

$$u_t = \sigma(W_u x_t + V_u h_{t-1} + b_u), \tag{1.22}$$

Reset gate:

$$r_t = \sigma(W_r x_t + V_r h_{t-1} + b_r), \tag{1.23}$$

Memory cell:

$$\tilde{h}_t = \tanh(W_h x_t + V_h(r_t * h_{t-1}) + b_r), \tag{1.24}$$

Output:

$$h_t = u_t h_{t-1} + (1 - u_t)\tilde{h}_t, \tag{1.25}$$

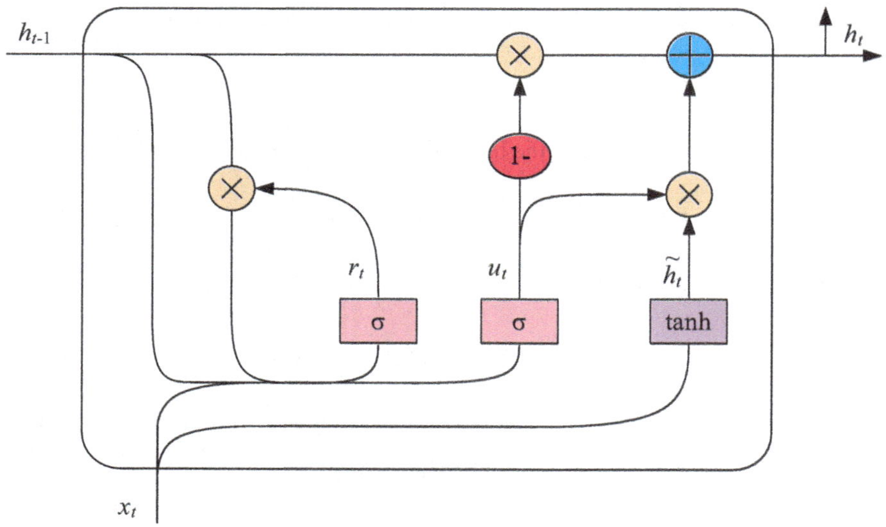

Fig. 1.8 The framework of GRU

where W_u, U_r, W_h, V_u, V_r, and V_h denote the weight matrix, b_i, b_f, b_c, and b_o denote the bias vector, and tanh denotes the activation function.

1.2.9 Bi-directional Gated Recurrent Unit (Bi-GRU)

A Bi-GRU is an improved GRU model. One-directional GRU needs to utilize previous information to deduce subsequent information, preventing it from accessing future context and integrating with contextual information, which may affect its prediction performance. A Bi-GRU adopts two GRU networks to train together, and starts their respective sequences from the converse ends while being connected back to the same output layer. Thus, it can integrate the past and future information of each point. A Bi-GRU consists of forward and backward calculation components. The horizontal direction represents the bi-directional flow of time sequence, and the vertical direction represents the one-directional flow from the input layer to the hidden layer, and then to the output layer. The specific framework of a Bi-GRU is shown in Fig. 1.9.

The forward calculation of hidden vector $\boldsymbol{h}$ is represented as follows:

$$\boldsymbol{h}_t = \mathrm{GRU}(x_t, \boldsymbol{h}_{t-1}) \tag{1.26}$$

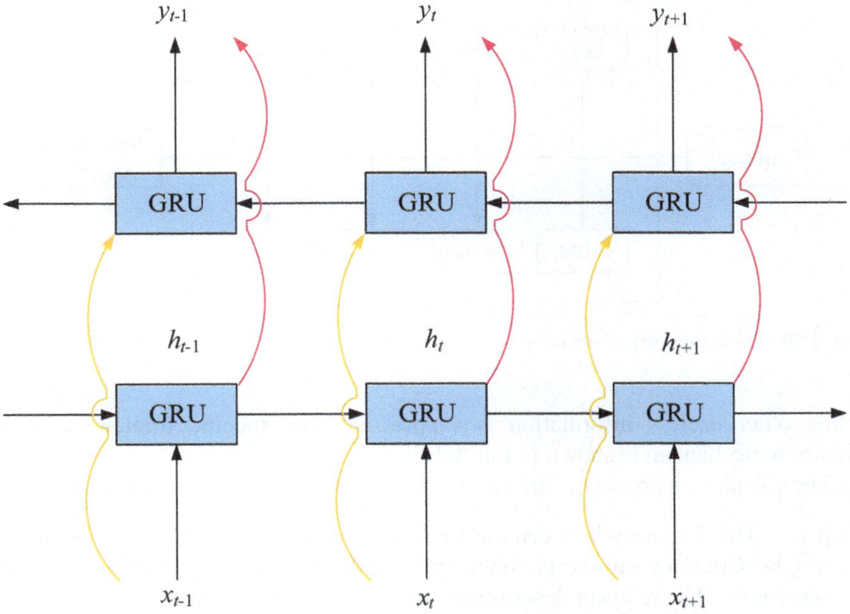

Fig. 1.9 The framework of Bi-GRU

The backward calculation of hidden vector $\overleftarrow{h}$ is described as follows:

$$\overleftarrow{h}_t = \text{GRU}(x_t, \overleftarrow{h}_{t-1}) \tag{1.27}$$

The output is described as follows:

$$y_t = g(W_{\boldsymbol{h}y}\boldsymbol{h}_t + W_{\overleftarrow{h}y}\overleftarrow{h}_t + b_y) \tag{1.28}$$

where x_t denotes the input data, y_t denotes the output in time t, $W_{\boldsymbol{h}y}$ and $W_{\overleftarrow{h}y}$ denote the weight matrix, and b_y denotes the bias term.

1.2.10 Attention

An attention mechanism imitates human visual processing (i.e., it aligns internal experience with external feelings to increase the observation precision of certain areas). For example, when browsing a picture, people first scan the global image quickly to obtain a target area (i.e., attention point) that requires the focus. However, more attention is then devoted to that point to obtain more detailed information

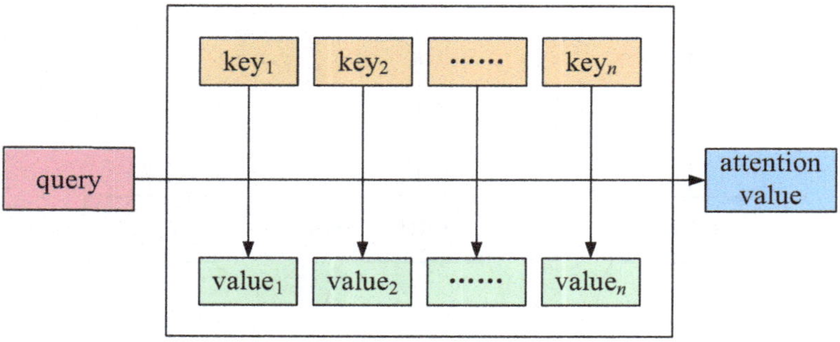

Fig. 1.10 The framework of attention

while other useless information is suppressed. The specific framework of an attention mechanism is shown in Fig. 1.10.

The calculation process is mainly divided into the following three steps.

Step 1: The similarity between the query and each key is calculated to obtain the weight. Common similarity functions include dot product, concatenating, and perceptron. The relevant descriptions are represented as follows:

$$f(Q, K) = \begin{cases} Q^T K \text{ , dot} \\ Q^T W_a K \text{ , general} \\ W_a[Q; K] \text{, concat} \\ v_a^t \tanh(W_a Q + U_a K) \text{ , perceptron} \end{cases} . \tag{1.29}$$

Step 2: Generally, a softmax function is applied to normalize these weights, which is represented as follows:

$$a_i = \text{softmax}(f(Q, K)) = \frac{\exp(f(Q, K_i))}{\sum_{j=1}^{n} \exp(f(Q, K_j)}. \tag{1.30}$$

Step 3: The final attention is obtained by weighting and summing the weight and corresponding key value, which is represented as follows:

$$\text{attention}\ (Q, K, V) = \sum_i a_i V_i. \tag{1.31}$$

At present, researchers tends to use identical keys and values in NLP.

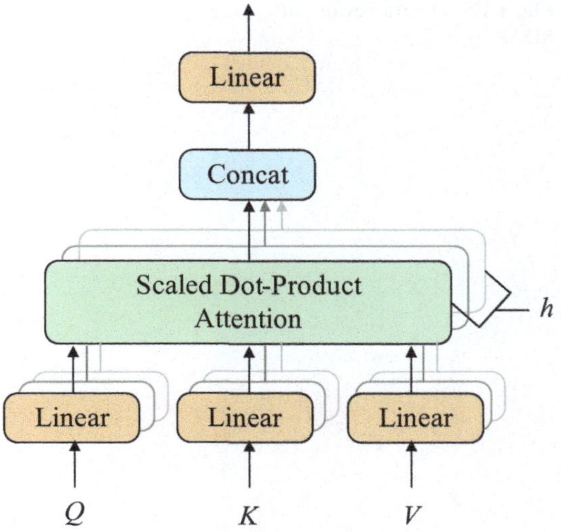

Fig. 1.11 The framework of MHA

1.2.11 Multi-Head Attention (MHA)

MHA is an attention mechanism variant that uses multiple queries to extract multiple groups of different information in parallel from input information for concatenating. The specific framework of a MHA is shown in Fig. 1.11.

First, a linear transformation is made for the query, key, and value. Then, they are input into the scaled dot product attention (SDA) mechanism, and the same operation is repeated h times. The input of each time is the linear transformation of the original input. The SDA framework is shown in Fig. 1.12.

An SDA is an attention mechanism of similarity calculation using a point product, a series of queries, a series of keys with dimensions d_k, and a series of values with dimensions d_v. The calculation process is represented as follows:

$$\text{attention}(Q, K, V) = \text{softmax}\left(\frac{QK^T}{\sqrt{d_k}}\right)V. \tag{1.32}$$

"Multi-head" denotes that a head is calculated each time, parameter W of the linear transformation for Q, K, and V is different each time, and the results of the h times SDA mechanism are concatenated. Then, they are conducted in a linear transformation once more to obtain a value, which is the MHA result. The calculation is represented as follows:

$$\text{multihead}(Q, K, V) = \text{concat}(head_1, head_2, \ldots, head_h)W^o, \tag{1.33}$$

$$head_i = \text{attention}(QW_i^Q, KW_i^K, VW_i^V), \tag{1.34}$$

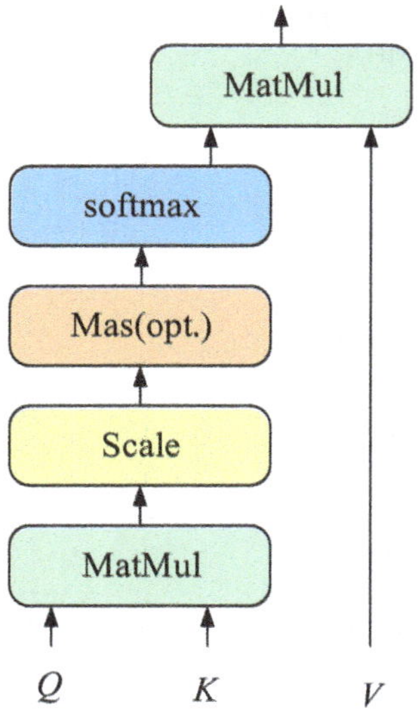

Fig. 1.12 The framework of SDA

where $W_i^Q \in \mathbb{R}^{d_k \bar{d}}$, $W_i^K \in \mathbb{R}^{d_k \bar{d}}$, $W_i^V \in \mathbb{R}^{d_V \bar{d}}$, and $W^o \in \mathbb{R}^{hd_V \bar{d}}$ denote a single attention function with d-dimensional keys, values, queries, and output, respectively.

1.2.12 Generative Adversarial Network (GAN)

GAN aims to estimate generative models via an adversarial process. Two models are simultaneously trained—one generative model (G), named generator, aims to capture the data distribution, and the other discriminative model (D), named discriminator, aims to estimate the probability that a sample comes from the training data rather than G. The most direct application of GAN is to configure both G and D as multi-layer perceptrons. The specific framework of a GAN is shown in Fig. 1.13.

To determine the distribution p_g of G over data x, it needs to define a prior on input noise $p_z(z)$, and then a mapping to data space can be represented as $G(z; \theta_g)$, where G is a differentiable function determined by the multi-layer perceptrons parameter represented by θ_g. In addition, a second multi-layer perceptron $D(x; \theta_d)$ is defined to output a single scalar, where $D(x)$ denotes the probability that x came from the data rather than p_g. D is trained to maximize the probability of assigning the correct label to both training examples and samples from G, and G is trained to minimize log(1-$D(G(z))$).

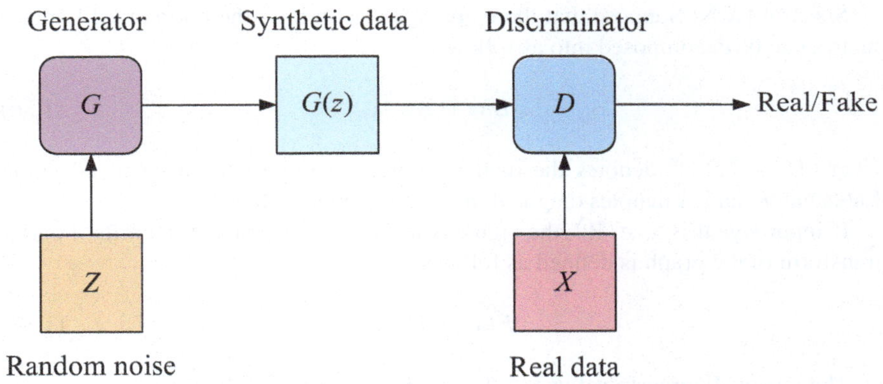

Fig. 1.13 The framework of GAN

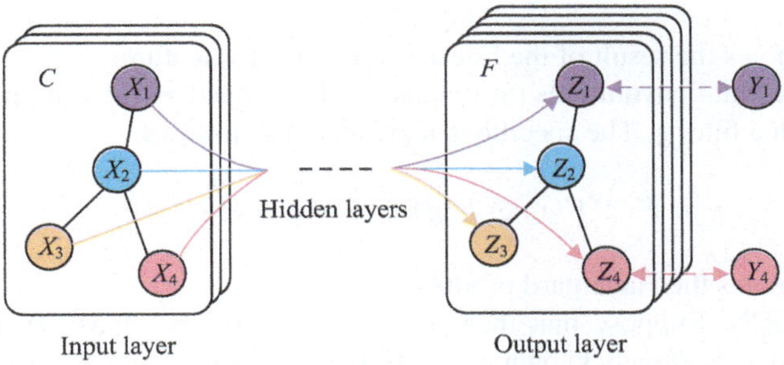

Fig. 1.14 The framework of GCN

Specifically, D and G play a two-player minimax game with a value function $V(G, D)$, the specific description is represented as follows:

$$\min_{G}\max_{D} V(D, G) = \mathbb{E}_{x\ p_{data}(x)}[\log D(x)] + \mathbb{E}_{z\ p_z(z)}[\log(1 - D(G(z)))]. \quad (1.35)$$

D aims to identify whether the data is the generated data or not as much as possible, and G aims to minimize the difference between the generated data and the real data. This means enhancing the discrimination ability of D and improving the ability G to generate data closer to real data during the loop process.

1.2.13 Graph Convolutional Network (GCN)

GCN contains two types of models—one is spectral GCN [29], and the other is spatial GCN [23]. The specific framework of a GCN is shown in Fig. 1.14.

Spectral GCN: Suppose that there are N nodes in a G, the normalized Laplace matrix can be decomposed into as follows:

$$L = U \Lambda U^T, \tag{1.36}$$

where $U \in \mathbb{R}^{N \times N}$ denotes the matrix of eigenvectors of the normalized graph Laplacian L, and Λ denotes diagonal matrix of its eigenvalues.

If input signal is $x \in \mathbb{R}^N$ the signal of each node is a scalar, then the Fourier transform of the graph is defined as follows:

$$F(x) = U^T x. \tag{1.37}$$

The inverse Fourier transform of the graph is defined as follows:

$$F^{-1}(\breve{x}) = U\breve{x}, \tag{1.38}$$

where $\breve{x}$ denotes the result of the Fourier transform for feature.

Thus, spectral convolutions on graphs can be defined as the multiplication of a signal x with a filter g. The specific is represented as follows:

$$x * g = F^{-1}(F(x) \odot F(g)) = U(U^T x \odot U^T g), \tag{1.39}$$

where $\odot$ denotes the Hadamard product.

Spatial GCN: Suppose that there is a graph transition matrix P. If the graph structure of data is already known, that is, if the similarity matrix S is already known, the transition matrix can be obtained. The specific is represented as follows:

$$P = D^{-1} S, \tag{1.40}$$

where D denotes a diagonal matrix, the absolute value of correlation matrix is utilized as similarity matrix S, and $D_{ii} = \sum_j S_{ij}$, $S_{ii} = \left|R_{ij}\right|$.

The transition expectation for k steps is represented as follows:

$$Q^{(0)} = I, Q^{(1)} = I + P, \ldots, Q^{(k)} = \sum\nolimits_{i=0}^{k} P^k, \tag{1.41}$$

where $[P^k]_{ij}$ and $Q_{ij}^{(k)}$ denote the transaction probability and expected number of visits, respectively, from node X_i to node X_j in k steps.

Each row for $Q^{(k)}$ can be utilized to obtain the nearest neighbors of a node. Thus, it can define the convolution over the node X_i by using the i-th row of $Q^{(k)}$. Suppose $\pi_i^{(k)}$ as the permutation order of the i-th row of $Q^{(k)}$ in a descend order, then the

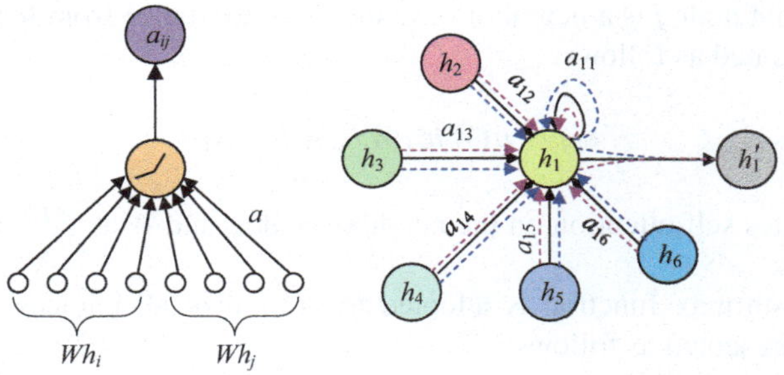

Fig. 1.15 The framework of GAT

ordered position between any two nodes can represents a global feature of all nodes. Thus, the inner product is represented as follows:

$$Conv(x) = \begin{bmatrix} x_{\pi_1^{(k)}(1)} & \cdots & x_{\pi_1^{(k)}(p)} \\ x_{\pi_2^{(k)}(2)} & \cdots & x_{\pi_2^{(k)}(p)} \\ \vdots & \ddots & \vdots \\ x_{\pi_N^{(k)}(N)} & \cdots & x_{\pi_N^{(k)}(p)} \end{bmatrix} \begin{bmatrix} w_1 \\ w_2 \\ \vdots \\ w_p \end{bmatrix}, \tag{1.42}$$

where *Conv* denotes the size p convolution over G, $x = (x_1, \ldots, x_N)^T \in \mathbb{R}^N$, weight $w \in \mathbb{R}^p$ is determined according to the distance deduced by the transition matrix.

1.2.14 Graph Attention Network (GAT)

Graph Attention Network (GAT) [53] is a network structure that combines graph convolution and attention mechanism. Based on GCN, it adopts attention mechanism to replace the inherent standardized function, assigns different weights to each neighbor node, without obtaining the information of graph structure in advance. The specific framework of a GAT is shown in Fig. 1.15.

Suppose that there are N number of nodes in G, h_i is a feature vector of each node, F is the number of features for each node. Thus, a set of node features h is represented as follows:

$$h = \{h_1, h_2, \ldots, h_N\}, h_i \in \mathbb{R}^F. \tag{1.43}$$

A new feature vector $h_i{}'$ is obtained by linearly transforming for h_i, and F' is the number of features for each node. Thus, a set of node features h' is represented as follows:

$$h' = \{h_1{}', h_2{}', \ldots, h_N{}'\}, h_i{}' \in \mathbb{R}^{F'}. \tag{1.44}$$

Suppose that node j is a neighbor of i, and the importance from feature of j for i can be represented as follows:

$$e_{ij} = a(Wh_i, Wh_j), j \in N_i, \tag{1.45}$$

where a denotes self-attention on the neighbor nodes, and $W \in \mathbb{R}^{F' \times F}$ denotes the weight matrix.

Then the softmax function is adopted to normalize all choices of j, and the specific is represented as follows:

$$a_{ij} = \text{sofimax}(e_{ij}) = \frac{\exp(e_{ij})}{\sum_{k \in N_i} \exp(e_{ij})}. \tag{1.46}$$

By concatenating h_i and h_j, and applying nonlinear activation function LeakyReLU, a_{ij} can be represented as follows:

$$a_{ij} = \frac{\exp(\text{LeakyReLU}(\boldsymbol{a}^T[Wh_i||Wh_j])}{\sum_{k \in N_i} \exp(\text{LeakyReLU}(\boldsymbol{a}^T[Wh_i||Wh_j])}, \tag{1.47}$$

where || denotes the concatenation operation.

The regularized attention coefficients of different nodes can be obtained through the above operations, which can be utilized to predict the output feature of each node. The specific is represented as follows:

$$h_i' = \sigma\left(\sum_{j \in N_i} a_{ij} Wh_j\right). \tag{1.48}$$

The above computing process is performed by using self-attention mechanism, and multi-head attention is also utilized by the process. If there are K independent attention mechanism adopted to execute the above transformation in Eq. (1.49), their feature vectors need to be concatenated together. The specific is represented as follows:

$$h_i' = \text{concat}\left(\sigma\left(\sum_{j \in N_i} a_{ij}^k W^k h_j\right)\right). \tag{1.49}$$

If multi-head attention is performed in the final layer of the network, the outputs of K attention mechanism are not concatenated, but averaged. The specific is represented as follows:

$$h_i' = \sigma\left(\frac{1}{K}\sum_{k=1}^{K}\sum_{j \in N_i} a_{ij}^k W^k h_j\right). \tag{1.50}$$

1.2.15 Performance Comparison of Each DL Model

The performance for different model are summarized in Table 1.1.

1.3 Structure of the Book

There are nine chapters in the book, and its structure is listed as follows.

In Chap. 1, we introduce relevant knowledge for TEC task, such as the emotion overview (i.e., emotion definition and classification) and TEC's applications, and also the overview of DL as the background for the whole book.

In Chap. 2, we present a new method for TCE task by using BERT and broad learning (BL). We first introduce the basic idea of bidirectional encoder representation from transformers (BERT) and BL, and the detailed implementation for TEC task. We adopt evaluation metrics like accuracy and macro-$F1$ to evaluate the performance of the proposed method. In addition, we conduct the experiment on two public data sets to verify the performance of the proposed BBL.

In Chap. 3, we present an approach for TCE task by using masked permuted pre-trained language model (MPNet) and cascading BL. We first introduce the basic idea of MPNet and cascading BL, and the detailed implementation for TEC task. To effectively maintain the performance of the cascading BL, we adopt the L-curve method to ensure balance between under-regularization and over-regularization in the regularization parameter optimization. Finally, we carry out extensive experiments on two public data sets to verify the effectiveness.

In Chap. 4, we present a new method for the TCE task by using dual BL. Firstly, texts are input into BERT to generate sentence embedding which is input into the first BL. Then, texts are input into MPNet to generate sentence embedding which is also input into the other BL. Finally, the concatenation is conducted to fuse the features from BERT and MPNet.

In Chap. 5, we present the single-source domain adaptation for emotion classification by integrating the advantages of CNN and BL. We first adopt CNN to extract domain-invariant and domain-specific features simultaneously, so as to train two more efficient classifiers by using BL. In addition, to take the advantage of these two classifiers, we design a co-training model to boost together for them.

In Chap. 6, we present the multi-source domain adaptation for emotion classification by using BL and DL. We first adopt BERT and Bi-LSTM to extract domain-invariant features from each source domain to the same target domain and then use BL to train multiple classifiers based on the domain-invariant features.

In Chap. 7, we present the emotion classification in textual conversations by using DL and BL. It aims to provide a better solution to capture the local contextual information (i.e., utterance-level) in an utterance, as well as the global contextual information (i.e., speaker-level) in a conversation on the basis of CNN, Bi-LSTM, and BL.

Table 1.1 Performance comparison for different model

No.	Name	Advantages	Disadvantages
1	DAN [28]	Its structure is simple, and it can extract richer data features.	It is prone to be gradient vanishing and gradient explosion.
2	DAE [54]	It can be trained directly by using raw data, and can be extended to large-scale data processing, which can be applied to various signal processing scenarios.	It needs a significant amount of training time and computing resources, and its performance is limited by the quality and quantity of training data.
3	CNN [30]	It is suitable for processing high-dimensional data and can automatically extract features.	It needs a large number of samples in training process, and does not consider the historical information. The gradient descent algorithm is prone to fall into local convergence for the training results.
4	RNN [20]	It is suitable for processing sequence information, with simple internal structure and low requirements for computing resources. Compared with LSTM and GRU, there are fewer parameters and better performance for dealing with short sequence tasks.	It is prone to long-distance dependency problem, and also be encountered gradient vanishing and gradient explosion problem in back propagation process.
5	LSTM [24]	It can better deal with the long-distance dependency problem, and can effectively reduce the risk of gradient vanishing or gradient explosion.	The ability for dealing with long-distance dependency is still not enough, the gradient vanishing problem can not be completely solved, and the training time is relatively long.
6	Bi-LSTM [48]	It can better capture long-distance dependency.	It still can not deal with long sequences well.
7	GRU [14]	It has simple structure, fast training speed, and easy convergence, and can alleviate the phenomenon of gradient vanishing or gradient explosion.	The gradient vanishing problem cannot be completely solved.
8	Bi-GRU [55]	It can obtain contextual information in texts, and with fast training speed.	It fails to deal with long sequences well.
9	Attention [27]	It has fewer parameters and lower time complexity and can obtain the global and local connection and can be utilized in parallel mode.	It is difficult to capture location information, that is, it is impossible to learn order relationship in the sequence.

(continued)

Table 1.1 (continued)

No.	Name	Advantages	Disadvantages
10	MHA [52]	It can obtain relevant information in different dimensions and sub-spaces, so as to achieve better performance. The ability for capturing long-distance features is better than that of single attention mechanism.	Compared with attention, there are many parameters and difficult to train for multi-head attention.
11	GAN [21]	GAN does not need to label data during training, and can generate diverse data samples.	Its interpretability is poor. The training process may be very difficult and unstable, sometimes leading to model collapse.
12	GCN [29]	Spectral GCN can make the classification problem easier. Spatial GCN can directly define convolution operation in spatial domain, which makes the model more intuitive.	Spectral GCN only be adapted to undirected graph, and can not be utilized for dynamic graph. Spatial GCN lacks of theoretical basis.
13	GAT [53]	It is suitable for directed graph and can assign different weights to each neighbor node. Because it is only related to adjacent nodes, there is no need to obtain the information of the whole graph.	Only neighbor information of 1-hop is exploited, while the information of multi-hop does not be utilized fully.

In Chap. 8, we present the emotion classification in textual conversations by combining the relational GAT (RGAT) and BL. It aims to provide a better solution to capture the contextual features (i.e., utterance-level) in an utterance, as well as the long-distance dependencies (i.e., speaker-level) in a conversation on the basis of Bi-LSTM, RGAT, and BL.

In Chap. 9, we provide an summary the book and the understanding of the landscape of emotion analysis, and hope in the near future to provide the possible reference to the potential readers.

References

1. Dictionary by merriam-webster, emotion, https://www.merriam-webster.com/dictionary/emotion
2. Development Report of China New Media Propagation of Microblog, forum and other we-media increase the threat on modern society, China Reading Newspaper (November 28 2012), http://epaper.gmw.cn/zhdsb/html/2012-11/28/nw.D110000zhdsb_20121128_3-18.htm
3. The 38th statistical report on the development of china's internet (August 2016), http://www.cnnic.net.cn/hlwfzyj/hlwxzbg/hlwt-jbg/20-1608/t20160803_54392.htm

4. Ahmad, Z., Jindal, R., Ekbal, A., Bhattachharyya, P.: Borrow from rich cousin: transfer learning for emotion detection using cross lingual embedding. Expert Systems With Applications **139**(112851), 1–12 (2020)
5. Alkouz, B., Aghbari, Z., Abawajy, J.: Tweetluenza: predicting flu trends from twitter data. Big Data Mining and Analytics **2**(4), 273–287 (2019)
6. Barbieri, F., Camacho-Collados, J., Ronzano, F., Espinosa-Anke, L., Ballesteros, M., Basile, V., Patti, V., Saggion, H.: Semeval 2018 task 2: Multilingual emoji prediction. In: Proceedings of the 12th International Workshop on Semantic Evaluation. pp. 24–33. New Orleans, Louisiana (June 2018)
7. Bharti, S., Vachha, B., Pradhan, R., Babu, K., Jena, S.: Sarcastic sentiment detection in tweets streamed in real time: a big data approach. Digital Communications and Networks **2**, 108–121 (2016)
8. Bouazizi, M., Ohtsuki, T.: Multi-class sentiment analysis on twitter: classification performance and challenges. Big Data Mining and Analytics **2**(3), 181–194 (2019)
9. Camachoa, D., Panizo-LLedot, A., Bello-Orgaz, G., Gonzalez-Pardo, A., Cambria, E.: The four dimensions of social network analysis: An overview of research methods, applications, and software tools. Information Fusion **63**, 88–120 (2020)
10. Cambria, E., Poria, S., Gelbukh, A., Thelwall, M.: Sentiment analysis is a big suitcase. IEEE Intelligent Systems **32**(6), 74–80 (2017)
11. Cao, L., Peng, S., Yin, P., Zhou, Y., Yang, A., Li, X.: A Survey of Emotion Analysis in Text Based on Deep Learning. In: Proceedings of the IEEE 8th International Conference on Smart City and Informatization (iSCI 2020). pp. 81–88. Guangzhou, China (2020)
12. Ceron, A., Curini, L., Iacus, S.M., Porro, G.: Every tweet counts? How sentiment analysis of social media can improve our knowledge of citizens political preferences with an application to Italy and France. New Media and Society **16**(2), 340–358 (2014)
13. Chen, H., Yin, C., R. Li, W.R., Xiong, Z., David, B.: Enhanced learning resource recommendation based on online learning style model. Tsinghua Science and Technology **25**(03), 348–356 (2020)
14. Cho, K., Merrienboer, B., Gulcehre, C., Bahdanau, D., Bougares, F., Schwenk, H., Bengio, Y.: Learning phrase representations using RNN encoder-decoder for statistical machine translation. In: Proceedings of the 2014 Conference on Empirical Methods in Natural Language Processing. pp. 1724–1734. Doha, Qatar (2014)
15. Choudhury, M.D., Counts, S., Horvitz, E.J., Hoff, A.: Characterizing and predicting postpartum depression from shared Facebook data. In: Proceedings of the 17th ACM Conference on Computer Supported Cooperative Work & Social Computing. pp. 626–638. Baltimore, MD, USA (2014)
16. Colneric, N., Demsar, J.: Emotion recognition on twitter: Comparative study and training a unison model. IEEE Transactions on Affective Computing **11**(3), 433–446 (2020)
17. Deng, J., Dong, W., Socher, R., Li, L., Li, K., Li, F.: ImageNet: a large-scale hierarchical image database. In: Proceedings of the 2009 IEEE Conference on Computer Vision and Pattern Recognition. pp. 248–255. Miami, FL, USA (June 2009)
18. Ekman, P.: An argument for basic emotions. Cognition and Emotion **6**(3/4), 169–200 (1992)
19. Fang, B., Jia, Y., Li, A., Yin, L.: Research progress and trend of cyberspace big search. Journal on Communications **36**(12), 1–8 (2015)
20. Giles, C.L., Kuhn, G.M., Williams, R.J.: Dynamic recurrent neural networks: Theory and applications. IEEE Transactions Neural Networks **5**, 153–156 (1994)
21. Goodfellow, I.J., Pouget-Abadie, J., Mirza, M., Xu, B., Warde-Farley, D., Ozair, S., Courville, A., Bengio, Y.: Generative adversarial nets. In: Proceedings of the Advances in Neural Information Processing Systems (NIPS 2014). pp. 2672–2680. Cambridge, MA, USA (2014)
22. Hao, Y., Zheng, Q., Chen, Y., Yan, C.: Recognition of abnormal behavior based on data of public opinion on the web. Journal of Computer Research and Development **53**(3), 611–620 (2016)
23. Hechtlinger, Y., Chakravarti, P., Qin, J.: A generalization of convolutional neural networks to graph-structured data (2017). https://doi.org/https://doi.org/10.48550/arXiv.1704.08165

24. Hochreiter, S., Schmidhuber, J.: Long short-term memory. Neural Computation **9**(8), 1735–1780 (1997)
25. Huang, X.: Introduction to Psychology. Beijing: Peoples Education Press (1991)
26. Hudlicka, E.: Guidelines for designing computational models of emotions. International Journal of Synthetic Emotions **2**(1), 26–79 (2011)
27. Itti, L., Koch, C.: Computational modelling of visual attention. Nature reviews neuroscience **2**(3), 194–203 (2001)
28. Iyyer, M., Manjunatha, V., Boyd-Graber, J., III, H.D.: Deep unordered composition rivals syntactic methods for text classification. In: Proceedings of the 53rd Annual Meeting of the Association for Computational Linguistics and the 7th International Joint Conference on Natural Language Processing. pp. 1681–1691. Beijing, China (July 2015)
29. Kipf, T.N., Welling, M.: Semi-supervised classification with graph convolutional networks. In: Proceedings of the 5th International Conference on Learning Representations (ICLR 2017) (2017)
30. LeCun, Y., Bottou, L., Bengio, Y., Haffner, P.: Gradient-based learning applied to document recognition. Browse Journals & Magazines **86**(11), 2278–2324 (1998)
31. Li, R., Lin, Z., Lin, H., Wang, W., Meng, D.: Text emotion analysis: a survey. Journal of Computer Research and Development **55**(1), 30–52 (2018)
32. Liao, J., Tang, J., Zhao, X.: Course drop-out prediction on mooc platform via clustering and tensor completion. Tsinghua Science and Technology **24**(4), 412–422 (2019)
33. Lin, C.: Emotional problems in socialist psychology. Science of Social Psychology **21**(83), 37–62 (2006)
34. Liu, B.: Sentiment Analysis: Mining Opinions, Sentiments, and Emotions. Cambridge University Press (2015)
35. Lv, Y., Wei, F., Cao, L., Peng, S., Niu, J., Yu, S., Wang, C.: Aspect-level sentiment analysis using context and aspect memory network. Neurocomputing **428**, 195–205 (2021)
36. Munezero, M., Montero, C., Sutinen, E., Pajunen, J.: Are they different? Affect, feeling, emotion, sentiment, and opinion detection in text. IEEE Transaction Affective Computing **5**(2), 101–111 (October 2014)
37. Parrott, W.: Emotions in social psychology: essential readings. Oxford, UK: Psychology Press (2001)
38. Paul, D., Li, F., Teja, M.K., Yu, X., Frost, R.: Compass: Spatio temporal sentiment analysis of US Election what Twitter says! In: Proceedings of the 23rd ACM International Conference on Knowledge Discovery and Data Mining. pp. 1585–1594. Halifax, Canada (2017)
39. Peng, S., Wang, G., Zhou, Y., Wan, C., Wang, C., Yu, S., Niu, J.: An immunization framework for social networks through big data based influence modeling. IEEE Transactions on Dependable and Secure Computing **16**(6), 984–995 (2019)
40. Peng, S., Zhou, Y., Cao, L., Yu, S., Niu, J., Jia, W.: Influence analysis in social networks: a survey. Journal of Network and Computer Applications **106**, 17–32 (2018)
41. Peng, S., Cao, L., Zhou, Y., Ouyang, Z., Yang, A., Li, X., Jia, W., Yu, S.: A survey on deep learning for textual emotion analysis in social networks. Digital Communications and Networks **8**, 745–762 (2022)
42. Peng, S., Zeng, R., Liu, H., Chen, G., Wu, R., Yang, A., Yu, S.: Emotion classification of text based on bert and broad learning system. In: Proceedings of The Asia Pacific Web (APWeb) and Web-Age Information Management (WAIM) Joint International Conference on Web and Big Data. pp. 382–396 (2021)
43. Plutchik, R.: The nature of emotions. Philosophical Studies **89**(4), 393–409 (October 2001)
44. Poria, S., Cambria, E., Bajpai, R., Hussain, A.: A review of affective computing: From unimodal analysis to multimodal fusion. Information Fusion **37**, 98–125 (2017)
45. Poria, S., Majumder, N., Mihalcea, R., Hovy, E.: Emotion recognition in conversation: research challenges, datasets, and recent advances. IEEE Access **7**, 100943–100953 (October 2019)
46. Salakhutdinov, R., Hinton, G.: Deep Boltzmann machines. In: Proceedings of the 12th International Conference on Artificial Intelligence and Statistics. pp. 448–455. Florida, USA (2009)

47. Saleem, T.J., Chishti, M.A.: Deep learning for the internet of things: potential benefits and use-cases. Digital Communications and Networks, https://doi.org/10.1016/j.dcan.2020.12.002. (2020)
48. Schuster, M., Paliwal, K.K.: Bidirectional recurrent neural networks. IEEE Transactions on Signal Processing **45**(11), 2673–2681 (1997)
49. Song, X., Jiang, W., Liu, X., Lu, H., Tian, Z., Du, X.: A survey of game theory as applied to social networks. Tsinghua Science and Technology **25**(6), 734–742 (2020)
50. Tucker, C., Pursel, B., Divinsky, A.: Mining student-generated textual data in MOOCs and quantifying their effects on student performance and learning outcomes. The ASEE Computers in Education (CoED) Journal **5**(4), 84 (2014)
51. Usama, M., Ahmad, B., Song, E., Hossain, M.S., Alrashoud, M., Muhammad, G.: Attention-based sentiment analysis using convolutional and recurrent neural network. Future Generation Computer Systems **113**, 571–578 (2020)
52. Vaswani, A., Shazeer, N., Parmar, N., Uszkoreit, J., Jones, L., Gomez, A., Kaiser, L., Polosukhin, I.: Attention is all you need. In: Proceedings of the 31st Conference on Neural Information Processing Systems. pp. 5999–6009. Long Beach, California, USA (2017)
53. Velickovic, P., Cucurull, G., Casanova, A., Romero, A., Lio, P., Bengio, Y.: Graph attention networks. In: Proceedings of the 6th International Conference on Learning Representations (ICLR 2018) (2018)
54. Vincent, P., Larochelle, H., Bengio, Y., Manzagol, P.A.: Extracting and composing robustfeatures with denoising autoencoders. In: Proceedings of the 25th International Conference on Machine Learning. pp. 1096–1103. USA (2008)
55. Wang, Z., Zhang, B.: Improved bi-gru model for imbalanced english toxic comments dataset. In: Proceedings of the 5th International Conference on Natural Language Processing and Information Retrieval. pp. 24–29 (2021)
56. Wu, X., Fu, L., Long, H., Yang, D., Lu, Y., Wang, X., Chen, G.: Adaptive diffusion of sensitive information in online social networks. IEEE Transactions on Knowledge and Data Engineering **33**(8), 3020–3034 (2021)
57. Xi, X., Zhou, G.: A survey on deep learning for natural language processing. Acta Automatica Sinica **42**(10), 1445–1465 (2016)
58. Xie, Z.: Modelling the dropout patterns of mooc learners. Tsinghua Science and Technology **25**(3), 313–324 (2020)
59. Yang, C., Lai, X., Hu, Z., Liu, Y., , Shen, P.: Depression tendency screening use text based emotional analysis technique. Journal of Physics: Conference Series **1237**, 1–10 (2019)
60. Zhang, J., Wang, Y., Yuan, Z., Jin, Q.: Personalized real-time movie recommendation system: practical prototype and evaluation. Tsinghua Science and Technology **25**(2), 180–191 (2020)
61. Zhang, L., Xu, C., Gao, Y., Han, Y., Du, X., Tian, Z.: Improved dota2 lineup recommendation model based on a bidirectional lstm. Tsinghua Science and Technology **25**(6), 712–720 (2020)
62. Zhang, P., Huang, X., Zhang, L.: Information mining and similarity computation for semi- / unstructured sentences from the social data. Digital Communications and Networks, https://doi.org/10.1016/j.dcan.2020.08.001. (2020)
63. Zhang, Z., Li, X., Gan, C.: Identifying influential nodes in social networks via community structure and influence distribution difference. Digital Communications and Networks **7**(1), 131–139 (2021)
64. Zhao, W., Guan, Z., Chen, L., He, X., Cai, D., Wang, B., Wang, Q.: Weakly-supervised deep embedding for product review sentiment analysis. IEEE Transactions on Knowledge and Data Engineering **30**(1), 185–197 (2018)
65. Zhou, M., Duan, N., Liu, S., Shum, H.Y.: Progress in neural NLP: modeling, learning, and reasoning. Engineering, doi: https://doi.org/10.1016/j.eng.2019.12.014 (2020)
66. Zhu, H., Shan, X., Hu, J.: 2011 China Internet public opinion analysis report (full text) (July 2012), http://yuqing.people.com.cn/n/2012/0727/c209170-18615551.html

Chapter 2
BERT and Broad Learning for Textual Emotion Classification

Abstract TEC is an important task in NLP. Most existing methods for TEC are based on DL. However, there are some weaknesses in DL, such as long training time and difficult convergence. Due to a flat network model of BL, which has the advantages of simple structure, incremental modeling and short training time, in this chapter, we explore to combine BERT and BL, named BBL, for TEC task. In addition, we conduct the relevant experiments on two public data sets to verify the performance of the proposed BBL.

2.1 Introduction

With the rapid development of social media and e-commerce, more and more users prone to publish their reviews contained real emotions in the social networks that may facilitate the advance of opinion mining and analysis [15, 39]. For example, Weibo[1] or Twitter[2] have become one of the most popular tools for people to express personal opinions and emotions [21]. Sentiment and emotion analysis have become two fundamental tasks for NLP, which were attracted by the academia [19].

In general, there are two main sub-tasks in textual emotion analysis: emotion recognition and emotion classification [2]. Emotion recognition aims to judge whether the text contains emotion by analyzing the text. Emotion classification aims to analyze the text containing emotion on the basis of emotion recognition and to predict the involving emotion towards a piece of text [18]. Some instances for TEC task are listed as follows:

E1. I especially **like** today's weather! (happy)
E2. So **scared**, every day so that. (fear)
E3. It's really **annoying**. I always get interrupted when I want to do something. (anger)

[1] https://weibo.com/.

[2] https://twitter.com/.

S. Peng, L. Cao, *Textual Emotion Classification Using Deep Broad Learning*, Socio-Affective Computing 11, https://doi.org/10.1007/978-3-031-67718-2_2

From the examples, it can be seen that these texts with emotional words such as "like", "scared" and "annoying" often show obvious emotional tendency. In the early research works, they mainly focus on studying bipartition (i.e., positive and negative) of sentiment classification, and based on tripartition (i.e., positive, negative, and neutral) of sentiment classification [29]. However, emotion classification based on bipartition and tripartition can not fully express the complex inner world of people, which ignores the subtle emotional changes, nor it can describe the user psychological state more comprehensively. Emotion classification is also called fine-grained sentiment classification [23]. Thus, sentiment classification can be regarded as bipartition-oriented or tripartition-oriented emotion classification. From the perspective of human psychology, emotion classification can characterize emotional attitude of people in multi-dimensional view.

Existing methods for TEC usually include the following three types: the methods based on emotion lexicon (EL) [33, 34, 40], traditional machine learning (ML) [27, 35, 44], and DL [1, 16, 36, 43].

The EL-based method mainly uses EL resources to extract the keywords of emotion expression and then classifies emotion from texts. For example, Xu et al. [40] constructed a Chinese emotion lexicon, called WordNet-Affect, based on the English lexicon WordNet [24]. Staiano and Guerini [34] built an emotion lexicon, called DepecheMood, with a crowd-sourced affective annotation manner. Song et al. [33] built a lexicon for emotion classification, called Lex-c, by combining seed words and emotions.

Salah et al. [30] constructed domain emotion dictionaries through two methods: (1) calculating the polarity and degree of words from the annotated corpus to generate a dictionary; (2) starting from the existing dictionaries—an extended domain dictionary based on classifier is learned by combining the semantic information, context information and relationship information of the domain corpus. Bernabé et al. [4] presented an approach to extract a polarity dictionary from the stock market. It tracks the price changes of particular stocks over time, functioning as a guiding polarity value and then linking with financial news about stocks.

However, the EL-based method takes a lot of time to construct and maintain a high-quality emotion dictionary. In addition, it heavily relies on the quality and coverage of EL. Because of the continuous emergence of new online words (e.g., stan, clapback), it is difficult to construct, maintain, and expand a high-quality EL. In addition, it ignores the contextual information in text, only simply to calculate the emotional intensity of text according to the number of emotional words, which may lead to a lack of feature.

Thus, some researchers began to explore to use ML for TEC task. The ML-based method often uses the supervised and semi-supervised learning algorithms, which rely on a large number of labeled data. Pang et al. [25] applied ML to classify emotion for the first time. Xin et al. [38] explored emotion intensities of words and phrases in an eight dimensional emotion space as features, and then adopted the support vector machine (SVM) to classify the emotion of text.

In addition, Zhao et al. [44] designed a classifier for sentiment analysis by using naive Bayes (NB). Purver and Battersby [27] proposed a semi-supervised emotion classification method by using SVMs. Suttles and Ide [35] designed a classifier for emotion classification by using NB. However, these supervised learning algorithms require a large number of high-quality data, and need more time and manpower to annotate the corpus, which may affect its practicality [28].

With the growing scale of data, there is difficulty in constructing a high-quality EL and in providing a large number of labeled data based on massive text data. EL-based method may take more massive time to be constructed and maintained. ML-based method needs more complex feature engineering [31], including text preprocessing (such as word frequency statistics) and feature selection (such as n-gram feature). In addition, the feature vectors trained by the ML-based algorithms are sparse and weak for word feature expression. Thus, these EL- and ML-based methods are difficult to be applied to conduct emotion classification. Thus, in recent years, many researchers have adopted DL models to conduct TEC task.

With the wide application of DL in NLP, more and more researchers adopt the DL models to solve the above problems. They utilize them to extract deep textual features. For example, CNN is one of the most popular DL models, which includes input layer, hidden layer and output layer and can automatically extract deep text features due to their characteristics of multiple layers and complex structures. Thus, it can solve the problems like complex feature engineering and difficult feature extraction.

Thus, in recent years, many DL-based methods are proposed for TEC task, and DL models are usually adopted to automatically extract features in texts. The DL-based methods can effectively overcome the shortcomings of the EL-based and ML-based methods to a certain extent. For example, Kim [33] and Zhang et al [44] utilized CNN, Abdul-Mageed and Ungar [27] and Tafreshi and Diab [35] exploited GRNN, and Basile et al. [3] and Xiao [37] employed Bi-LSTM and attention, to solve TEC task. However, its training time is long and convergence speed is pretty slow, due to its complex structure and too many parameters.

Although DL models have strong learning capability, due to a large number of parameters and complex network structure, they lead to long training time, large amounts of computing resources, and difficulty in theoretical analysis. In addition, DL models such as RNN and LSTM are prone to gradient explosion and gradient vanishing during their training [26].

BL is a flat network model, which has been proposed by Chen and Liu [8]. Instead of stacking and expanding neurons in a deep manner, BL expands neurons including feature nodes and enhancement nodes in a wide manner. Then, the output weight is calculated by the pseudo inverse. Compared with the common DL models, BL has the advantages of simple network structure, short training time, and strong generalization ability. It is an effective alternative method for DL, and its training process can be extended to the incremental learning model without retraining when new nodes are added. Chen et al. [9] also proved theoretically the universal approximation ability of BL. This method has achieved better results in many applications, such as regression [13], hyperspectral image classification [10], visual recognition [14], semi-supervised learning [22], target location [45], emotion recognition [42], automatic control [20], and fault diagnosis [41].

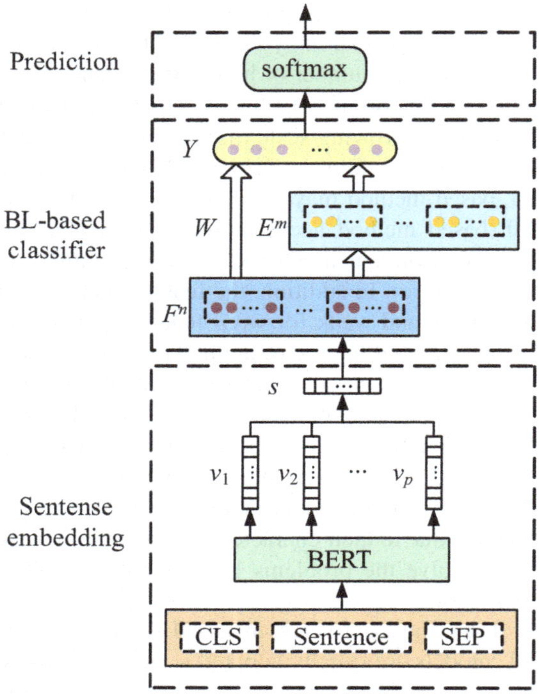

Fig. 2.1 The framework of BBL

2.2 Methodology

In this section, we will introduce the modeling process of BBL, which includes three layers like encoder, classifier, and prediction. The overall structure of BBL is shown in Fig. 2.1.

Firstly, the text is input into the pre-training model BERT to obtain the contextual semantic related word embeddings, and the word embeddings are averaged mathematically to obtain the sentence embedding of the text. The above operation is repeated for each sentence. Secondly, the feature nodes of BL are generated by linear transformation of sentence embeddings, and then the enhancement nodes of BL are generated by nonlinear transformation of feature nodes. Then, feature nodes and enhancement nodes are input into the output layer of BL. Finally, the probability of emotion of each text can be predicted by using the softmax function.

2.2.1 Encoding Layer

BERT [11] is a new general language model that provides the support for different NLP tasks. On the basis of the transformer architecture released by Google in 2017,

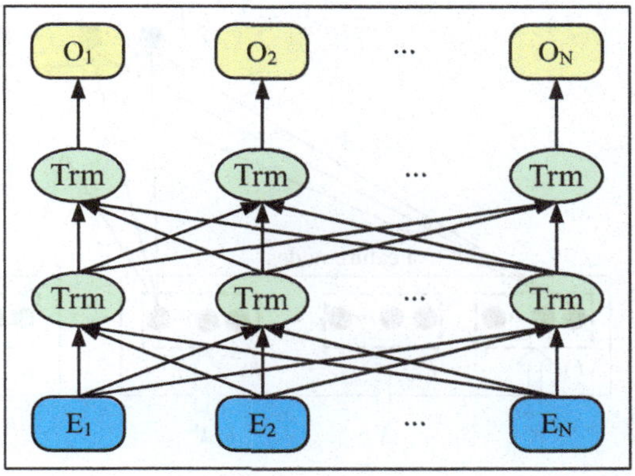

Fig. 2.2 The structure of BERT

it trains the depth of pre-training bidirectional representation by jointly adjusting bidirectional transformers in all layers. Its advantages lie in the use of new pre-training tasks of masked language model (MLM) and next sense prediction (NSP). The structure of BERT is shown in Fig. 2.2

BERT is composed of multiple transformer blocks. The encoder of transformer consists of a multi-head attention layer and a feed-forward layer, which performs layer normalization for each layer. Multi-head attention layer contains multiple self-attention layers that can obtain current word information and semantic information. Multi-head attention layer expands the capacity by obtaining different aspects of attention information from multiple self-attention layers.

The input of BERT consists of three parts: i) Token embedding: it uses a specific number to represent the corresponding word in sentences according to the relationship between a word and its number in vocabulary; ii) Segment embedding: it uses 0 and 1 to distinguish any two sentences; iii) Position embedding: it uses a sequence number to indicate the position of a word in text.

To conduct the vectorized representation of sentences, each of them in a data set needs to be encoded as a fixed length vector as an embedding. We use the last hidden layer of BERT to obtain the vector representation of each word in each sentence, and then average the vector of each word to obtain the sentence embedding. It is supposed that $v_i \in \mathbb{R}^{768}$ denotes the weight of the i-th word in a sentence S. The sentence embedding of S is represented as follows:

$$S = \frac{1}{p}\sum_{i=1}^{p} v_i \in \mathbb{R}^{768}, \tag{2.1}$$

where p denotes the total number of words in a sentence S, and v_i denotes the word vector of the i-th word in a sentence.

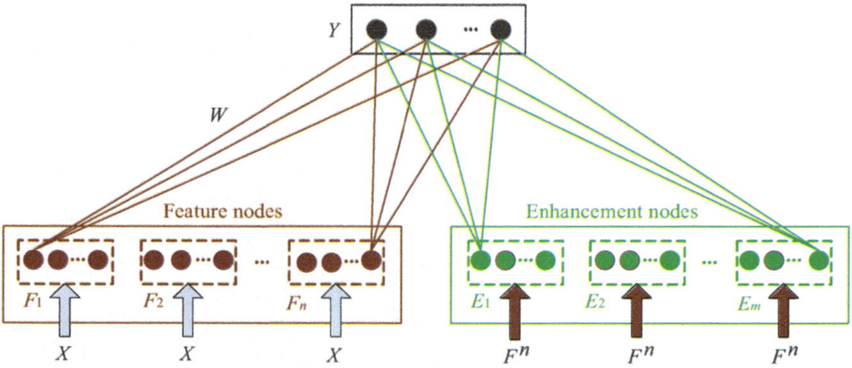

Fig. 2.3 The framework of BL

2.2.2 *BL-Based Classifier*

The basic idea of BL is that sentence embeddings are mapped into feature nodes for extracting main information of text, and then feature nodes are transformed into enhancement nodes to further extract the useful information of text. To be specific, sentence embeddings are mapped into high dimensional linear feature space, and then these linear features are mapped into high dimensional nonlinear feature space through nonlinear activation function. Finally, both features extracted from feature nodes and enhancement nodes are input into the output layer to obtain the probability distribution of emotions. To solve the output weight, we transform the solution of pseudo inverse matrix into the problem of ridge regression. The framework of BL is shown in Fig. 2.3.

Given a training set $\{S, Y\} \in \mathbb{R}^{N\times(M+C)}$, where N denotes the total number of samples, M denotes the dimension of word\sentence embedding S, and C denotes the total number of class. S is linearly mapped into n groups of feature nodes. The i-th group of feature nodes F_i is represented as follows:

$$F_i = g\left(SW_{fi} + \beta_{fi}\right) \in \mathbb{R}^{N\times q}, i = 1, 2, \ldots, n, \tag{2.2}$$

where g denotes the linear activation function, q denotes the number of feature nodes in each group, and W_{fi} and β_{fi} denote the weight matrix and bias matrix generated randomly, respectively.

Let $F^n \triangleq [F_1, F_2, \ldots, F_n]$ denote the matrix of n groups of feature nodes, which represents all linear features of text data. F^n are nonlinearly mapped into m groups of enhancement nodes. The j-th group of enhancement nodes E_j is represented as follows:

$$E_j = \varphi\left(F^n W_{ej} + \beta_{ej}\right) \in \mathbb{R}^{N\times r}, j = 1, 2, \ldots, m, \tag{2.3}$$

where φ denotes nonlinear activation function (e.g., sigmoid, tanh, ReLU), r denotes the total number of enhancement nodes in each group, and W_{ej} and β_{ej} denote the weight matrix and bias matrix generated randomly, respectively.

Let $E^m \triangleq [E_1, E_2, \ldots, E_m]$ denote the matrix of m groups of enhancement nodes, which represents all the nonlinear features of text. The matrix concatenated by F^n and E^m is utilized to generate the actual input, which can be represented as follows:

$$A = \left[F^n, E^m\right] \in \mathbb{R}^{N\times(nq+mr)}. \tag{2.4}$$

Thus, the final estimated output $\hat{Y}$ can be represented as follows:

$$\begin{aligned}\hat{Y} &= \left[g\left(SW_{f1} + \beta_{f1}\right), \ldots, g\left(SW_{fn} + \beta_{fn}\right)\right.\\ &\quad \left.| \varphi\left(F^n W_{e1} + \beta_{e1}\right), \ldots, \varphi\left(F^n W_{em} + \beta_{em}\right)\right] W\\ &= \left[F_1, F_2, \ldots, F_n \,|E_1, E_2, \ldots, E_m\right] W\\ &= \left[F^n, E^m\right] W\\ &= AW.\end{aligned} \tag{2.5}$$

The output weights W can be represented as follows:

$$W = A^+ Y, \tag{2.6}$$

where A^+ denotes the pseudo inverse matrix of A.

However, calculating the pseudo inverse matrix directly will take a large amounts of time, and may lead to over-fitting. Thus, the ridge regression is adopted as objective function to solve W, which can be represented as follows:

$$\underset{W}{\operatorname{argmin}} \left(\|Y - AW\|_2^2 + \lambda \|W\|_2^2 \right), \tag{2.7}$$

where λ denotes regularization parameters.

According to the regularized least square method, W can be represented as follows:

$$W = (\lambda I + AA^{\mathrm{T}})^{-1} A^{\mathrm{T}} Y. \tag{2.8}$$

It is notable that the weights of middle layers W_{fi}, W_{ej}, β_{fi}, and β_{ej} are fixed, and only the output weights W need to be updated, which is different from the end-to-end learning manner of deep learning. The learning process of BL is listed in Algorithm 2.1.

Algorithm 2.1 The learning process of BL

Input: Sentence embeddings S
Output: Output weights W

```
for i = 1; i ≤ n do
    Random W_fi, β_fi
    Compute F_i = g(S W_fi + β_fi)
end for
Set the feature nodes F^n ≜ [F_1, F_2, ..., F_n]
for j = 1; j ≤ m do
    Random W_ej, β_ej
    Compute E_j = φ(F^n W_ej + β_ej)
end for
Set the enhancement nodes E^m ≜ [E_1, E_2, ..., E_m]
Set A = [F^n, E^m]and compute output weights W
```

2.3 Experiment and Analysis

In this section, we will provide the detailed description of the evaluation metrics, data set, baseline methods, parameter setting, and provide the comparison analysis among our proposed method with the baseline methods. In addition, we verify the efficiency of BBL through the comparative experiments, and analyze the impact on our proposed method from different pre-training language models. The experimental device is a laptop with Intel-i5-9400 2.9 GHz CPU and 16 GB memory, and server with Nvidia V100 GPU.

2.3.1 Evaluation Metrics

TEC task aims to classify those reviews into several emotional types like happy, sad, angry, fear, surprise, and neutral. Thus, this task is called a multi-classification problem. It can be transformed into a problem of binary classifications as well. In this book, we adopt two main metrics (i.e., accuracy and Macro-$F1$) [5] to evaluate the performance of our proposed models.

Accuracy refers to the ratio of the correctly classified samples to the total number of samples, which can be represented as follows:

$$\text{Accuracy} = \frac{\sum_{i=1}^{C}(TP_i + TN_i)}{\sum_{i=1}^{C}(TP_i + FP_i + TN_i + FN_i)}, \tag{2.9}$$

where TP_i (True Positive) denotes the number of the positive samples which are correctly classified, TN_i (True Negative) denotes the number of the negative samples which are falsely classified, FP_i (False Positive) denotes the number of the positive samples which are falsely classified, and FN_i (False Negative) denotes the number of the negative samples which are correctly classified.

Because the classes of data set applied in the experiment are imbalanced, which may cause inaccuracy to the true performance of the model well. Thus, Macro-$F1$ is also adopted as the second evaluation metric, which is defined as the average value of $F1$ values of all class, where $F1$ value is the harmonic average of precision and recall.

Precision (P) refers to the fraction of the texts correctly classified, for a given sentiment, over the total number of tweets classified as belonging to that sentiment. It can be represented as follows:

$$P_i = \frac{TP_i}{TP_i + FP_i}, \tag{2.10}$$

where P_i denotes precision.

Recall (R) refers to the fraction of texts correctly classified, for a given sentiment, over the total number of tweets actually belonging to that sentiment. It can be represented as follows:

$$R_i = \frac{TP_i}{TP_i + FN_i}, \tag{2.11}$$

where R_i denotes recall.

According to the value of precision, recall, we can use Eqs. (2.10) and (2.11) to compute Macro-$F1$, which can objectively reflect the performance of the model under class imbalance. It can be represented as follows:

$$F1_i = \frac{2 \times P_i \times R_i}{P_i + R_i}, \tag{2.12}$$

$$\text{Macro-}F1 = \frac{1}{C}\sum_{i=1}^{C} F1_i, \tag{2.13}$$

where $F1_i$ denotes $F1$ values of the i-th emotion class.

2.3.2 Data Set

In our experiments, we carry out comparative experiments on the English text data set from SemEval-2019 Task 3: EmoContext contextual emotion detection in the text, which is provided by Chatterjee et al. [7], and the Chinese text data set from the evaluation of Weibo emotion classification technology in the ninth China national

Table 2.1 Statistics of the SemEval

Emotion	Train	Validation	Test
Happy	5191	180	369
Sad	6357	151	308
Angry	6027	182	324
Other	12585	2242	4508

Table 2.2 Statistics of the SMP2020-EWECT usual

Emotion	Train	Validation	Test
Happy	5379	391	1018
Sad	4990	346	900
Angry	8344	586	1508
Fear	1220	87	210
Surprise	2086	170	373
Neutral	5749	420	990

Table 2.3 Statistics of the SMP2020-EWECT virus

Emotion	Train	Validation	Test
Happy	4423	923	1540
Sad	649	165	219
Angry	1322	314	463
Fear	555	75	190
Surprise	197	17	68
Neutral	1460	476	520

conference on social media processing (SMP2020-EWECT),[3] which includes two kinds of data: usual and virus. The distribution of each class of emotion is shown in Tables 2.1, 2.2, and 2.3, respectively.

2.3.3 Baseline Method

To verify the effectiveness of our proposed method in TEC, we compare it with the following baseline methods:

- **SVM** [32]: It is a commonly used baseline model for text classification tasks, which classify two classes by optimal decision boundary. In experiment, "one-versus-one" is adopted to train a classifier for every two classes.
- **TextCNN** [17]: It is a text classification model by using multiple convolution kernels with different sizes to combine automatically the n-gram features, so as to obtain deep-level semantic information.

[3] https://smp2020ewect.github.io/.

- **RCNN** [12]: It is a text classification model which adopts the bidirectional RNN to extract contextual features and uses the maximum pooling layer to capture key features.
- **Bi-LSTM**: It is the baseline model for most natural language processing tasks. Sentence is treated as a sequence and input into Bi-LSTM to extract contextual features. It also captures the long-distance dependencies of text.
- **Bi-LSTM-ATTN** [46]: It adds an attention layer for the Bi-LSTM. The attention mechanism can help to capture key information of text.

2.3.4 Parameter Setting

For a fair comparison,these DL-based models for TEC task (e.g., TextCNN, RCNN, Bi-LSTM, Bi-LSTM-ATTN) adopt the word embedding generated by BERT. The word embedding with dimension of 768 is adopted as the input for them. In SemEval data set, three rounds of dialogue are connected by the token "[SEP]" and input into BERT. We set sufficient epochs and introduce the early stopping [6] to ensure that the model can converge to a suitable position. A proper learning rate is applied to avoid model falling into a local optimum at early stage. Thus, optimization strategy of learning rate is introduced to obtain well-performed model. These methods based on DL are implemented in tensorflow framework, and their hyperparameter settings are shown in Table 2.4.

In addition, BBL and SVM take sentence embedding generated by BERT with dimension of 768. The sentence embedding is the output of the pooling layer in BERT. As to SemEval, the three sentence embeddings corresponding to three dialogues are concatenated as the input of the model. We adjust the hyperparameters of proposed model according to the actual performance of the validation set, and then determine the best set of parameters for testing. The hyperparameters to be

Table 2.4 Hyperparameter settings for the training of relevant DL models

Model	Parameter	Value
TextCNN	Kernel size	3/4/5
	Activation function	ReLU
RCNN	Bi-LSTM layer	1
	Bi-LSTM unit	400
	Kernel size	2
	Activation function	ReLU
Bi-LSTM	Bi-LSTM layer	2
	Bi-LSTM unit	400
Bi-LSTM-ATTN	Bi-LSTM layer	1
	Bi-LSTM unit	400
	Attention layer	1

Table 2.5 Relevant parameters for BBL

Parameter	SemEval	Usual	Virus
(n, k)	(41,50)	(23,50)	(10,50)
(m, q)	(41,50)	(25,50)	(15,50)
λ	0.0005	0.0003	0.0003
Activation function	tanh	tanh	tanh

Table 2.6 Accuracy of each model

Model	SemEval	Usual	Virus
SVM	82.69	70.08	67.84
TextCNN	**89.53**	76.04	76.79
RCNN	89.39	74.26	71.87
Bi-LSTM	87.55	76.10	77.07
Bi-LSTM-ATTN	87.46	**76.56**	**78.00**
BBL	89.00	76.51	76.38

Table 2.7 Macro-F1 of each model

Model	SemEval	Usual	Virus
SVM	58.19	61.68	51.94
TextCNN	73.61	72.92	61.00
RCNN	63.52	69.41	56.26
Bi-LSTM	68.47	72.34	60.80
Bi-LSTM-ATTN	72.57	72.52	**61.85**
BBL	**73.86**	**72.54**	58.48

adjusted mainly include the number of feature nodes and enhancement nodes, and regularization parameters, which are shown in Table 2.5.

2.3.5 *Main Result*

The accuracy of the baseline methods and our proposed method on three kinds of data sets are shown in Table 2.6.

From Table 2.6, it is known that TextCNN outperforms the other models on SemEval, and Bi-LSTM-ATTN outperforms the other models on Usual and Virus. As to the SemEval, apart from the TextCNN and RCNN, BBL outperforms the other baseline models. As to the Usual, BBL is lower than Bi-LSTM-ATTN by 0.04, but it outperforms the other baseline models. As to Virus, BBL is lower than TextCNN, Bi-LSTM, and Bi-LSTM-ATTN by 0.41, 0.69, and 1.62, respectively. The main reason is that the scale of Virus is small, and the learning of small-scale data is also a challenge for BBL. In addition, it can be seen that BBL can improve TEC performance more effectively than that of SVM.

The Macro-$F1$ of the baseline methods and our proposed method on three kinds of data sets are shown in Table 2.7.

From Table 2.7, it can be observed that TextCNN and Bi-LSTM-ATTN perform best among all of the baseline methods. This shows the effectiveness of CNN and attention mechanism for feature extraction of text. As to the SemEval and Usual, BBL achieves the best Macro-$F1$ of 73.86 and 72.54, respectively. As to the Virus, BBL achieves the Macro-$F1$ of 58.48, which is slightly lower than that of TextCNN, Bi-LSTM, and Bi-LSTM-ATTN. The main reasons are that BL is difficult to extract the deep semantic information of text and capture the contextual feature.

To explain this difference in performance, it is important to understand the nature of these models. TextCNN and Bi-LSTM-ATTN both fit to conduct emotion classification. TextCNN can extract local features of different sizes by setting different sizes of convolution kernel, which can make the feature vectors extracted from the network model more diverse and representative. Through adding an attention layer to the output of hidden layer, Bi-LSTM-ATTN can not only obtain contextual semantic information, but also calculate the relationship between each word and all words in a text. However, RCNN applies region proposal that selects several regions in data for training, which may ignore some crucial semantic information.

2.4 Summary

In this chapter, we explored BL for TEC task and proposed a novel method based on BERT and BL. For the given evaluation metrics accuracy and macro-$F1$, we conducted many experiments on the public data sets like SemEval-2019 Task 3 and SMP2020-EWECT. The experimental results show that BL and BERT can improve the performance, even better than many classical deep learning models. However, the advantage of general BL does not seem to be very remarkable.

References

1. Abdul-Mageed, M., Ungar, L.: EmoNet: Fine-grained emotion detection with gated recurrent neural networks. In: Proceedings of the 55th Annual Meeting of the Association for Computational Linguistics (Volume 1: Long Papers). pp. 718–728. Vancouver, Canada (Jul 2017)
2. Aman, S., Szpakowicz, S.: Identifying expressions of emotion in text. In: Proceedings of the International Conference on Text, Speech and Dialogue. pp. 196–205 (2007)
3. Basile, A., Franco-Salvador, M., Pawar, N., Stajner, S., Rios, M.C., Benajiba, Y.: Symantoresearch at semeval-2019 task 3: Combined neural models for emotion classification in human-chatbot conversations. In: Proceedings of the 13th International Workshop on Semantic Evaluation. pp. 330–334. Minneapolis, Minnesota, USA (2019)
4. Bernabe-Moreno, J., Tejeda-Lorente, A., Herce-Zelaya, J., Porcel, C., Herrera-Viedma, E.: A context-aware embeddings supported method to extract a fuzzy sentiment polarity dictionary. Knowledge-Based Systems **190**, 105236 (2020)
5. Cao, Z., Zhou, Y., Yang, A., Peng, S.: Deep transfer learning mechanism for fine-grained cross-domain sentiment classification. Connection Science **33**(4), 911–928 (2021)

6. Caruana, R., Lawrence, S., Giles, L.: Overfitting in neural nets: Backpropagation, conjugate gradient, and early stopping. Advances in neural information processing systems pp. 402–408 (2001)
7. Chatterjee, A., Narahari, K.N., Joshi, M., Agrawal, P.: Semeval-2019 task 3: Emocontext: contextual emotion detection in text. In: Proceedings of the 13th International Workshop on Semantic Evaluation. pp. 39–48. Minneapolis, Minnesota (2019)
8. Chen, C.P., Liu, Z.: Broad learning system: An effective and efficient incremental learning system without the need for deep architecture. IEEE transactions on neural networks and learning systems **29**(1), 10–24 (2018)
9. Chen, C.P., Liu, Z., Feng, S.: Universal approximation capability of broad learning system and its structural variations. IEEE transactions on neural networks and learning systems **30**(4), 1191–1204 (2019)
10. Chu, Y., Lin, H., Yang, L., Sun, S., Diao, Y., Min, C., Fan, X., Shen, C.: Hyperspectral image classification with discriminative manifold broad learning system. Neurocomputing **442**, 236–248 (2021)
11. Devlin, J., Chang, M.W., Lee, K., Toutanova, K.: BERT: Pre-training of deep bidirectional transformers for language understanding. In: Proceedings of the 2019 Conference of the North American Chapter of the Association for Computational Linguistics: Human Language Technologies, Volume 1 (Long and Short Papers). pp. 4171–4186. Minneapolis, Minnesota (Jun 2019)
12. Girshick, R., Donahue, J., Darrell, T., Malik, J.: Rich feature hierarchies for accurate object detection and semantic segmentation. In: Proceedings of the IEEE Conference on Computer Vision and Pattern Recognition (CVPR). pp. 580–587. Columbus, Ohio (June 2014)
13. Jin, J.W., Chen, C.P.: Regularized robust broad learning system for uncertain data modeling. Neurocomputing **322**, 58–69 (2018)
14. Jin, J., Li, Y., Yang, T., Zhao, L., Duan, J., Chen, C.P.: Discriminative group-sparsity constrained broad learning system for visual recognition. Information Sciences pp. 1–41 (2021)
15. Khatua, A., Khatua, A., Cambria, E.: Predicting political sentiments of voters from twitter in multi-party contexts. Applied Soft Computing **97**, 106743 (2020)
16. Kim, Y.: Convolutional neural networks for sentence classification. In: Proceedings of the 2014 Conference on Empirical Methods in Natural Language Processing (EMNLP). pp. 1746–1751. Doha, Qatar (Oct 2014)
17. Kim, Y.: Convolutional neural networks for sentence classification. In: Proceedings of the 2014 Conference on Empirical Methods in Natural Language Processing (EMNLP). pp. 1746–1751. Doha, Qatar (Oct 2014)
18. Li, S., Xu, J., Zhang, D., Zhou, G.: Two-view label propagation to semi-supervised reader emotion classification. In: Proceedings of the 26th International Conference on Computational Linguistics. pp. 2647–2655. Osaka, Japan (2016)
19. Li, Z., Xie, H., Cheng, G., Li, Q.: Word-level emotion distribution with two schemas for short text emotion classification. Knowledge-Based Systems **227**, 107163 (2021)
20. Lin, J., Liu, Z., Chen, C.P., Zhang, Y.: Three-domain fuzzy wavelet broad learning system for tremor estimation. Knowledge-Based Systems **192**, 105295 (2020)
21. Liu, B., Tang, S., Sun, X., Chen, Q., Cao, J., Luo, J., Zhao, S.: Context-aware social media user sentiment analysis. Tsinghua Science and Technology **25**(4), 528–541 (2020)
22. Liu, Z., Huang, S., Jin, W., Mu, Y.: Broad learning system for semi-supervised learning. Neurocomputing **444**, 38–47 (2021)
23. Lv, Y., Wei, F., Cao, L., Peng, S., Niu, J., Yu, S., Wang, C.: Aspect-level sentiment analysis using context and aspect memory network. Neurocomputing **428**, 195–205 (2021)
24. Miller, G.A.: Wordnet: a lexical database for english. Communications of the ACM **478**(11), 39–41 (1995)
25. Pang, B., Lee, L., Vaithyanathan, S.: Thumbs up? sentiment classification using machine learning techniques. In: Proceedings of the 2002 Conference on Empirical Methods in Natural Language Processing (EMNLP 2002). pp. 79–86 (2002)

26. Pascanu, R., Mikolov, T., Bengio, Y.: On the difficulty of training recurrent neural networks. In: Proceedings of the 30th International Conference on Machine Learning
27. Purver, M., Battersby, S.: Experimenting with distant supervision for emotion classification. In: Proceedings of the 13th Conference of the European Chapter of the Association for Computational Linguistics. pp. 482–491 (2012)
28. Ran, L., Zheng, L., Hailun, L., Weiping, W., Dan, M.: Text emotion analysis: A survey. Journal of Computer Research and Development **55**(1), 30–52 (2018)
29. Ran, L., Zheng, L., Hailun, L., Weiping, W., Dan, M.: Text emotion analysis: A survey. Journal of Computer Research and Development **55**(1), 30–52 (2018)
30. Salah, Z., Coenen, F., Grossi, D.: Generating domain-specific sentiment lexicons for opinion mining. In: Advanced Data Mining and Applications. pp. 13–24. Berlin, Heidelberg (2013)
31. Scott, S., Matwin, S.: Feature engineering for text classification. In: Proceedings of the 16th International Conference on Machine Learning. pp. 379–388 (1999)
32. Shoesmith, E.: Estimation of dependences based on empirical data. Journal of the Royal Statistical Society: Series D (The Statistician) **33**(3), 324–324 (1984)
33. Song, K., Feng, S., Gao, W., Wang, D., Chen, L., Zhang, C.: Build emotion lexicon from microblogs by combining effects of seed words and emoticons in a heterogeneous graph. In: Proceedings of the 26th ACM Conference on Hypertext and Social Media. pp. 283–292 (2015)
34. Staiano, J., Guerini, M.: Depechemood: A lexicon for emotion analysis from crowd-annotated news. In: Proceedings of the 52nd Annual Meeting of the Association for Computational Linguistics. pp. 427–433 (2014)
35. Suttles, J., Ide, N.: Distant supervision for emotion classification with discrete binary values. In: Proceedings of the 14th International Conference on Intelligent Text Processing and Computational Linguistics. pp. 121–136 (2013)
36. Tafreshi, S., Diab, M.: Emotion detection and classification in a multigenre corpus with joint multi-task deep learning. In: Proceedings of the 27th International Conference on Computational Linguistics (COLING 2018). pp. 2905–2913. Santa Fe, New Mexico, USA (Aug 2018)
37. Xiao, J.: Figure eight at semeval-2019 task 3: Ensemble of transfer learning methods for contextual emotion detection. In: Proceedings of the 13th International Workshop on Semantic Evaluation. pp. 220–224. Minneapolis, Minnesota, USA (2019)
38. Xin, K., F, R., Wu, Y.: Bottom up: Exploring word emotions for Chinese sentence chief sentiment classification. In: Proceedings of the 6th International Conference on Natural Language Processing and Knowledge Engineering (NLPKE 2010). pp. 1–5 (2010)
39. Xing, F., Malandri, L., Zhang, Y., Cambria, E.: Financial sentiment analysis: An investigation into common mistakes and silver bullets. In: Proceedings of the 28th International Conference on Computational Linguistics. pp. 978–987. Barcelona, Spain (Online) (2020)
40. Xu, J., Xu, R., Zheng, Y., Q. Lu, K.F., Wong, Wang, X.: Chinese emotion lexicon developing via multi-lingual lexical resources integration. In: Proceedings of the 14th International Conference on Intelligent Text Processing and Computational Linguistics. pp. 174–182 (2013)
41. Yu, W., Zhao, C.: Broad convolutional neural network based industrial process fault diagnosis with incremental learning capability. IEEE Transactions on Industrial Electronics **67**(6), 5081–5091 (2019)
42. Zhang, T., Wang, X., Xu, X., Chen, C.L.P.: Gcb-net: Graph convolutional broad network and its application in emotion recognition. IEEE Transactions on Affective Computing **14**(8), 1–10 (2019)
43. Zhang, Y., Fu, J., She, D., Zhang, Y., Wang, S., Yang, J.: Text emotion distribution learning via multi-task convolutional neural network. In: Proceedings of the 27th International Joint Conference on Artificial Intelligence (IJCAI 2018). pp. 4595–4601 (2018)
44. Zhao, J., Dong, L., Wu, J., Xu, K.: Moodlens: An emoticon-based sentiment analysis system for chinese tweets. In: Proceedings of the 18th ACM SIGKDD international conference on Knowledge discovery and data mining. pp. 1528–1531 (2012)

45. Zhong, D., Liu, F.: Rf-osfbls: An rfid reader-fault-adaptive localization system based on online sequential fuzzy broad learning system. Neurocomputing **390**, 28–39 (2020)
46. Zhou, P., Shi, W., Tian, J., Qi, Z., Li, B., Hao, H., Xu, B.: Attention-based bidirectional long short-term memory networks for relation classification. In: Proceedings of the 54th Annual Meeting of the Association for Computational Linguistics. pp. 207–212. Berlin, Germany (2016)

Chapter 3
Cascading Broad Learning for Textual Emotion Classification

Abstract Although many DL-based methods are proposed for TEC task, they suffer from certain disadvantages such as long training time and difficult convergence. Motivated by the above challenges, and to improve the performance of general BL, in this chapter, we propose a new method for this task, named CBL, which is based on the cascading BL and MPNet. Texts are input into the MPNet to generate sentence embedding to better obtain semantic information. Cascading BL is utilized to improve the ability of feature extraction of text by cascading the feature nodes and enhancement nodes simultaneously in general BL. In addition, the L-curve method is applied to ensure balance between under-regularization and over-regularization in the regularization parameter optimization. Extensive experiments are conducted on two public data sets to verify the effectiveness.

3.1 Introduction

In recent years, most existing models [3] for TEC task are based on DL. On the basis of CNN, Kulshreshtha et al. [7] proposed a neural architecture LE-PC-DNN for emotion intensity detection based on CNN, which combines the convolutional layers of CNN with fully connected layers in a non-sequential or 'parallel' fashion to improve system performance. Zhang et al. [20] proposed a multi-task CNN for textual emotion analysis, based on emotion distribution learning. On the basis of LSTM, Mohammadi et al. [11] proposed a neural feature extraction for contextual emotion detection. The model utilizes an attention-based recurrent neural network, and conducts experiment with GloVe and ELMo embeddings, alongside the POS tags as input, LSTM and GRU as recurrent units, and a neural or an SVM classifier. Li et al. [8] presented an emotion classification method for the short text, based on the skip-gram model and LSTM.

On the basis of the gated recurrent neural networks (GRNN), Abdul-Mageed and Ungar [1] proposed a method for the fine-grained emotion detection using the gated recurrent neural networks. Tafreshi and Diab [16] proposed a joint multi-task learning model by using a GRNN, and trained it with a multigenre emotion corpus to predict emotions for four type of genres. On the basis of the attention

S. Peng, L. Cao, *Textual Emotion Classification Using Deep Broad Learning*,
Socio-Affective Computing 11, https://doi.org/10.1007/978-3-031-67718-2_3

mechanism, Yu et al. [19] proposed a transfer learning architecture to transfer the sentiment classification knowledge to emotion classification tasks, via a dual attention mechanism. Yang et al. [18] proposed an framework, called interpretable relevant emotion ranking with event-driven attention (IRER-EA), based on RNN and attention mechanism.

On the basis of Bi-LSTM, Ahmad et al. [2] built a DL model for the emotion detection of Hindi language. They adopted CNN, Bi-LSTM, cross-lingual embeddings, and different transfer learning strategies for their purpose. He and Xia [5] presented a joint binary neural network (JBNN) model for multi-label emotion classification. In addition, Zhou et al. [21] proposed an emotional network (EmNet) for multi-label emotion classification, which jointly learns emotion lexicons and conducts classification. It consists of three components, such as sentence encoder, emotion generator, and emotion classifier. Ishiwatari et al. [6] proposed a relational position encoding method based on the relational graph attention networks to recognize human emotions in textual conversation.

These DL-based methods usually adopt a DNN containing multiple hidden layers to automatically extract features of text, which can effectively overcome the shortcomings of the methods EL- and ML-based to a certain extent. However, as to the DL-based methods, the training time is long and convergence speed is pretty low because of the complex structure and too many parameters in DL.

Although BL has the advantages of simple structure, incremental modelling, convenient theoretical analysis and strong generalization ability, it is difficult to capture contextual information of text due to its feed-forward network structure. In addition, pre-training language model like BERT has achieved the state-of-the-art performance on TEC task [21]. However, BERT neglects dependency among predicted tokens.

To address above problems, we adopt the cascading BL and MPNet [14] for TEC task. Specifically, we first use MPNet to generate sentence embeddings. It not only can enable sentence embeddings to capture the semantic-enriched and context-enriched information, but also can reduce the distance between sentence embeddings with same emotion. Then, we use cascading BL to improve the feature extraction capability of general BL through cascading the feature nodes and cascading enhancement nodes, respectively. At last, we employ the L-curve method for regularization parameter optimization to ensure that our proposed model can keep balance between under-regularization and over-regularization, so as to improve the performance of TEC.

3.2 Methodology

In this section, we will describe emotion classification model based on cascading BL and MPNet. Firstly, the overall framework of our proposed model is provided, and then each part of the model is described in detail, including MPNet-based sentence embeddings, cascading structure for BL, and L-curve parameter optimization for selecting regularization parameter.

3.2.1 Framework of CBL

The framework of our proposed method for TEC task is shown in Fig. 3.1.

The main idea of our proposed method is described as follows: The texts are input into MPNet to obtain sentence embeddings v with dimension of 768, which capture the relevant emotion information. Then, v is mapped into feature nodes F^n and enhancement nodes E^m to obtain high dimensional linear features and non-linear features, respectively. In addition, both features are utilized as the actual input and input into the output layer Y. At last, the softmax is adopted to obtain the probability distribution of emotions.

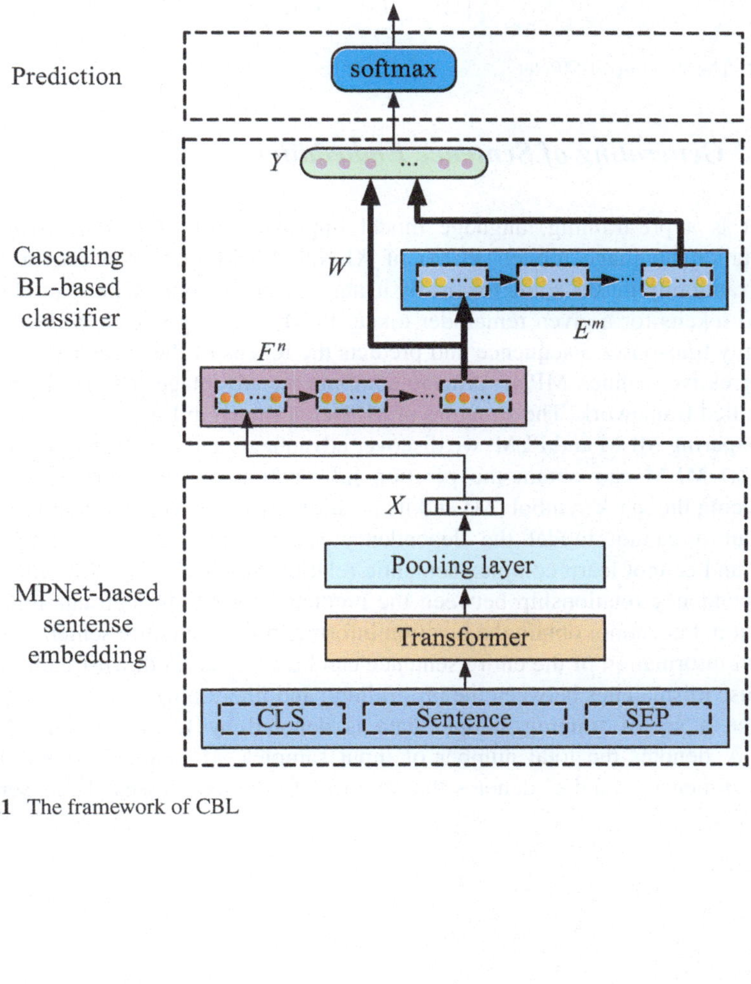

Fig. 3.1 The framework of CBL

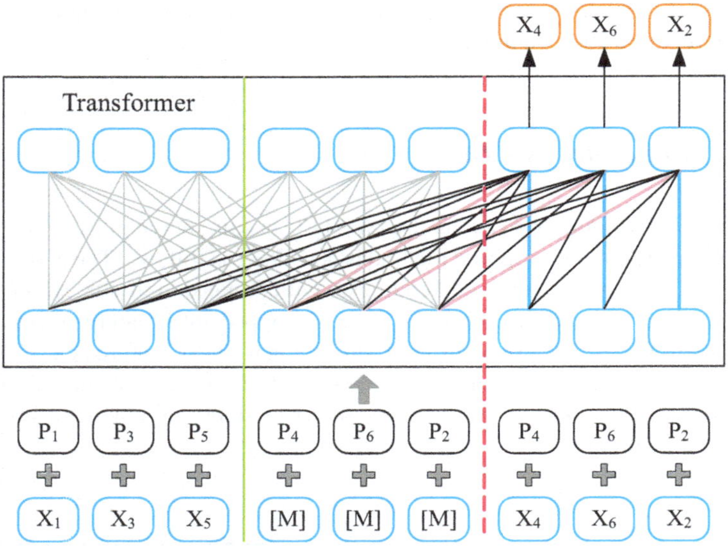

Fig. 3.2 The structure of MPNet

3.2.2 Generating of Sentence Embeddings

MPNet is a pre-training language model improved from MLM of BERT and permutation language model (PLM) of XLNet. MLM was proposed in BERT, which randomly masks some tokens by using a mask symbol [M] and predicts the masked tokens for a given remainder token. PLM was proposed in XLNet, which randomly transposes a sequence and predicts the tokens of the predicted part in an autoregressive manner. MPNet aims to combine the advantages of MLM and PLM in a unified framework. The structure of MPNet is shown in Fig. 3.2

Comparing MLM and PLM, we discover advantages and disadvantages between them: (i) MLM can obtain the position information of the entire sentence by introducing the mask symbol [M] and its position information in the non-predictive part, but it cannot model the dependency relationship between the predicted tokens and cannot learn complex semantic relationships well; (ii) PLM can model the dependency relationship between the predicted tokens through autoregressive prediction, but cannot obtain the position information of the entire sentence. As the position information of the entire sentence can be obtained in downstream tasks, it can cause mismatches between the pre-training and fine-tuning.

Given a set of training sample data is denoted by $\{X, Y\} \in \mathbb{R}^{N\times(M+C)}$, where N denotes the total number of input samples, M denotes the number of feature dimension, and C denotes the number of emotion classes. Thus, sentence embedding S is generated via MPNet, which are represented as follows:

$$S = \text{MPNet}(X, \theta) \in \mathbb{R}^{Nl \times 768}, \tag{3.1}$$

where θ denotes the relevant parameters, and l denotes the length of text.

3.2.3 *Cascading Broad Learning*

Since general BL is a feed-forward network of single-layer mapping, it only generates feature nodes and enhancement nodes through one layer of linear mapping and nonlinear mapping, which makes it difficult to capture high-level abstract features for TEC task. To solve the above problem, we use the cascading BL to conduct the multi-layer mapping of features to improve the ability of feature extraction.

The basic idea of cascading BL is that the sentence embeddings are linearly mapped into multiple groups of feature nodes in cascading order, and then all the feature nodes are nonlinearly mapped into multiple groups of enhancement nodes in cascading order. Finally, both features extracted from feature and enhancement nodes are input into the output layer to obtain the probability distribution of emotions. The solution of output weight is the same as the general BL. The structure of cascading BL is shown in Fig. 3.3.

Given a training set $\{S, Y\} \in \mathbb{R}^{N \times (M+C)}$, S is linearly mapped into the first group of feature nodes F_1, which can be represented as follows:

$$F_1 = g\left(SW_{f1} + \beta_{f1}\right) \triangleq g^1\left(S; \left\{W_{f1}, \beta_{f1}\right\}\right) \in \mathbb{R}^{N \times q}, \tag{3.2}$$

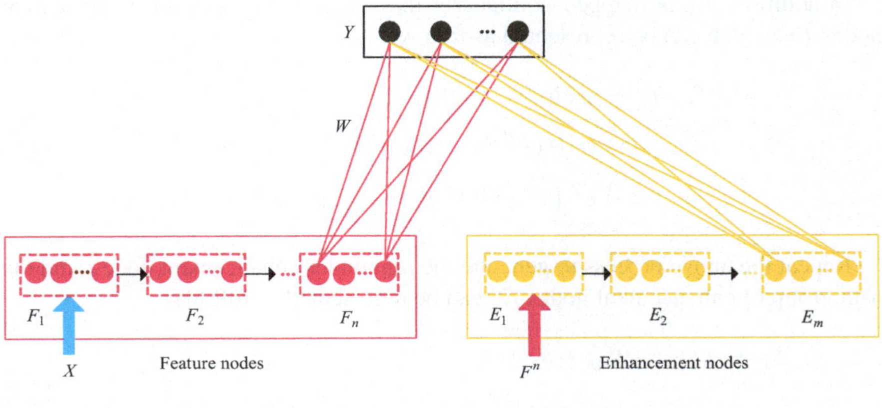

Fig. 3.3 The structure of cascading BL

where g denotes the linear activation function, q denotes the number of feature nodes in each group, and W_{f1} and β_{f1} denote the weight matrix and bias matrix generated randomly, respectively.

Then, feature nodes F_1 are linearly mapped into the second group of feature nodes F_2, which can be represented as follows:

$$\begin{aligned} F_2 &= g\left(F_1 W_{f2} + \beta_{f2}\right) \\ &= g\left(g\left(S W_{f1} + \beta_{f1}\right) W_{f2} + \beta_{f2}\right) \\ &\triangleq g^2\left(S; \left\{W_{fi}, \beta_{fi}\right\}_{i=1,2}\right) \in \mathbb{R}^{N \times q}. \end{aligned} \tag{3.3}$$

Repeat the above process to generate the remainder feature nodes, and the k-th group of feature nodes F_k can be represented as follows:

$$\begin{aligned} F_k &= g\left(F_{k-1} W_{fk} + \beta_{fk}\right) \\ &\triangleq g^k\left(S; \left\{W_{fi}, \beta_{fi}\right\}_{i=1}^{k}\right) \in \mathbb{R}^{N \times q}, k = 2, 3, \ldots, n, \end{aligned} \tag{3.4}$$

where n denotes the number of group of feature nodes, and W_{fk} and β_{fk} denote the weight matrix and bias matrix generated randomly, respectively.

Thus, $F^n \triangleq [F_1, F_2, \ldots, F_n]$ is concatenated by n groups of feature nodes. It is mapped nonlinearly into the first group of enhancement nodes E_1, which can be represented as follows:

$$E_1 = \varphi\left(F^n W_{e1} + \beta_{e1}\right) \triangleq \varphi^1\left(F^n; \{W_{e1}, \beta_{e1}\}\right) \in \mathbb{R}^{N \times r}, \tag{3.5}$$

where r denotes the total number of enhancement nodes in each group, and W_{e1} and β_{e1} denote weight matrix and bias matrix generated randomly, respectively.

In addition, E_1 is mapped nonlinearly into the second group of enhancement nodes E_2, which can be represented as follows:

$$\begin{aligned} E_2 &= \varphi\left(E_1 W_{e2} + \beta_{e2}\right) \\ &= \varphi\left(\varphi\left(F^n W_{e1} + \beta_{e1}\right) W_{e2} + \beta_{e2}\right) \\ &\triangleq \varphi^2\left(F^n; \left\{W_{ej}, \beta_{ej}\right\}_{j=1,2}\right) \in \mathbb{R}^{N \times r}. \end{aligned} \tag{3.6}$$

Repeat the above process to generate the remainder enhancement nodes, and the u-th group of enhancement nodes E_u can be represented as follows:

$$\begin{aligned} E_u &= \varphi\left(E_{u-1} W_{eu} + \beta_{eu}\right) \\ &\triangleq \varphi^u\left(F^n; \left\{W_{ej}, \beta_{ej}\right\}_{j=1}^{u}\right) \in \mathbb{R}^{N \times r}, u = 2, 3, \ldots, m, \end{aligned} \tag{3.7}$$

where m denotes the number of group of enhancement nodes, and W_{eu} and β_{eu} denote the weight matrix and bias matrix generated randomly, respectively.

Thus, $E^m \triangleq [E_1, E_2, \ldots, E_m]$ is concatenated by m groups of enhancement nodes. Then, F^n and E^m are concatenated to construct $A = [F^n, E^m]$ as the actual input. Thus, the final estimated output of cascading BL $\hat{Y}$ can be represented as follows:

$$\begin{aligned}\hat{Y} &= \Big[g^1\left(S; \{W_{f1}, \beta_{f1}\}\right), \ldots, g^n\left(S; \{W_{fi}, \beta_{fi}\}_{i=1}^{n}\right) \\ &\quad \mid \varphi^1\left(F^n; \{W_{e1}, \beta_{e1}\}\right), \ldots, \varphi^m\left(F^n; \{W_{ej}, \beta_{ej}\}_{j=1}^{m}\right)\Big] W \\ &= [F_1, F_2, \ldots, F_n \mid E_1, E_2, \ldots, E_m]\, W \\ &= \left[F^n, E^m\right] W \\ &= AW,\end{aligned} \tag{3.8}$$

where W is calculated by Eq. (2.8) in the Chap. 2.

The learning process of cascading BL is listed in Algorithm 3.1.

Algorithm 3.1 The learning process of cascading BL

Input: Sample data X
Output: The weights W

Random W_{f1}, β_{f1}
Compute $F_1 = g\left(SW_{f1} + \beta_{f1}\right)$
for $k = 2; k \leq n$ **do**
  Random W_{fk}, β_{fk}
  Compute $F_k = g\left(F_{k-1}W_{fk} + \beta_{fk}\right)$
end for
Set the feature nodes $F^n \triangleq [F_1, F_2, \ldots, F_n]$
Random W_{e1}, β_{e1}
Compute $E_1 = \varphi\left(F^n W_{e1} + \beta_{e1}\right)$
for $u = 2; j \leq m$ **do**
  Random W_{eu}, β_{eu}
  Compute $E_u = \varphi\left(E_{u-1}W_{eu} + \beta_{eu}\right)$
end for
Set the enhancement nodes $E^m \triangleq [E_1, E_2, \ldots, E_m]$
Set $A = [F^n, E^m]$and compute the output weights W

3.2.4 *Selection of Regularization Parameter*

Since regularization can prevent over-fitting effectively and can improve the generalization ability of model, it has been widely adopted in ML. As shown in $\underset{W}{\operatorname{argmin}}\left(\|Y - AW\|_2^2 + \lambda \|W\|_2^2\right)$ mentioned in Chap. 2, the objective function of

BL is ridge regression, which prevent the least squares falling into global optimal solution by adding penalty term $\lambda \parallel W \parallel_2^2$.

However, the estimation results of ridge regression are quite different under different regularization parameters. If λ is set too small, it will be prone to overfitting because of large variance and small bias. If λ is set too large, the estimation error will be large due to the large bias and small variance. Thus, it is essential to select a suitable λ to ensure the classification performance of the cascading BL.

The L-curve [4] is an effective solution for the regularization parameter optimization. First, the norm of residuals $\|Y - AW\|_2^2$ and the norm of regularized solution $\|W\|_2^2$ are calculated under different λ. Then, all of the data points $\left(\log\left(\|Y - AW\|_2^2\right), \log\left(\|W\|_2^2\right)\right)$ are fitted with a smooth decreasing curve, leading to an obvious inflection point to connect the part of large slope and the part of small slope. Its overall shape is similar to "L". Finally, the value corresponding to the inflection point is selected as the final result for λ.

To be specific, at the stage of large slope, slight increase in $\|Y - AW\|_2^2$ will make $\|W\|_2^2$ drop significantly, which will cause the model to fall into the stage of under-regularization. In contrast, increase in $\|W\|_2^2$ will make $\|Y - AW\|_2^2$ drop significantly at the stage of small slope, which will cause the model to fall into the stage of over-regularization. The parameter at the inflection point can better maintain a balance between them.

In the experiment, it is assumed that the distribution of validation set is the same as the test set. L-curve is adopted to select the best regularization parameter and then verify it on the test set. The process of L-curve parameter optimization is listed in Algorithm 3.2.

Algorithm 3.2 The process of L-curve parameter optimization

Input: Regularization parameters $\lambda_n, i = 1, 2, \ldots, n$
Output: The best value λ

1: **for** $i = 1; i \leq n$ **do**
2: Train classifier α_i with λ_i and calculate $\|W\|_2^2$
3: Validate classifier α_i and calculate $\|Y - AW\|_2^2$
4: **end for**
5: Fit points $\left(\log\left(\|Y - AW\|_2^2\right), \log\left(\|W\|_2^2\right)\right)$
6: Select the most appropriate value near the inflection point as λ

3.3 Experiment and Analysis

In this section, we will provide the description of the data set, baseline model, parameter setting, and the comparison analysis among our proposed method with baseline methods. In addition, we conduct some experiments to further verify the efficiency of CBL, and to analyze the impact of our proposed method from different pre-training language models. The experimental device is a laptop with Intel-i5-9400 2.9 GHz CPU and 16 GB memory, and server with Nvidia V100 GPU.

3.3.1 Data Set

To verify the effectiveness of our proposed method, we conduct comparative experiments on the data sets from SemEval-2019 Task 3 (SemEval), SMP2020-EWECT Usual (Usual), and SMP2020-EWECT Virus (Virus). Detailed information for each kind of data set is shown in Chap. 2.

Since the SemEval data set is collected from online text, it contains many emoticons, informal words and irregular grammar, etc., which are unfavourable to the generating of sentence embedding. Thus, we adopt the Ekphrasis toolkit [15] to preprocess the text as follows: (i) Emoticons are converted into the corresponding emotion words (e.g.,":-("is converted into "Unhappyface"); (ii) Informal words are transformed into corresponding correct words (e.g.,"Thks" is converted into "Thanks"); and (iii) Capital letters are converted into lowercase letters. In addition, we also use it to process the data set Usual and Virus due to much invalid information as follows: (i) Remove the url, email and "@" token; (ii) Traditional Chinese character and capital letters are converted to simplified Chinese character and lowercase letters respectively; and (iii) Emoticons are converted into the corresponding emotional words.

3.3.2 Baseline Method

To verify the effectiveness of CBL for TEC task, apart from the baseline models (i.e., SVM, TextCNN, RCNN, Bi-LSTM, and Bi-LSTM-ATTN) mentioned by the Chap. 2, we also compare it with the following baseline methods:

- **ERNIE-CNN**:[1] It is an emotion classification model provided in SMP2020-EWECT. The pre-training language model ERNIE is utilized as the embedding layer and a convolutional neural network is utilized for feature extraction.
- **ERNIE-RNN**: It is similar to the ERNIE-CNN, the difference is that recurrent neural network is utilized to extract contextual feature.
- **YUN-HPCC** [13]: It is ranked the 10th method of SemEval-2019 Task 3 proposed by Li and Zhang [13]. It achieves a Micro-$F1$ score of about 75.88% on the final evaluation.

[1] https://github.com/smp2020ewect/smp2020ewect.github.io/tree/master/reports.

- **SCIA** [15]: It improves performance by the ensemble with four different models: contextual Bi-LSTM, categorical Bi-LSTM, binary convolutional Bi-LSTM, and GRU.
- **DLCF-DCA** [17]: It consists of DLCF and DCA, where DLFCF aims to capture local context based on max distance and DCA aims to pay attention to the cluster which is important for classification.
- **ASC-pos-wt** [9]: It is a mechanism to capture positional bias, namely positional-biased weight, which can be applied to BERT for enhancing the performance of classification.

3.3.3 Parameter Setting

In addition, BL, CBL and SVM all take sentence embedding generated by MPNet with dimension of 768. As to SemEval, the three sentence embeddings corresponding to three dialogues are concatenated as the input of model. We adjust the relevant hyperparameters of our proposed model according to the performance of validation set, and finally select the best set of parameters for testing. The hyperparameters that need to be adjusted mainly include the number of feature nodes, the number of enhancement nodes number, and regularization parameters. We take SemEval as an example to show how to select the corresponding hyperparameters of CBL.

L-curve is adopted to select proper regularization parameter λ. Set λ [0.000005, 0.00001, . . . , 0.1] for CBL. The corresponding L-curve is drawn on the validation set, which is shown in Fig. 3.4.

From Fig. 3.4, it can be observed that the value near the inflection point is between 0.005 and 0.0005, and the model is at the equilibrium between under

Fig. 3.4 The L-curve of CBL

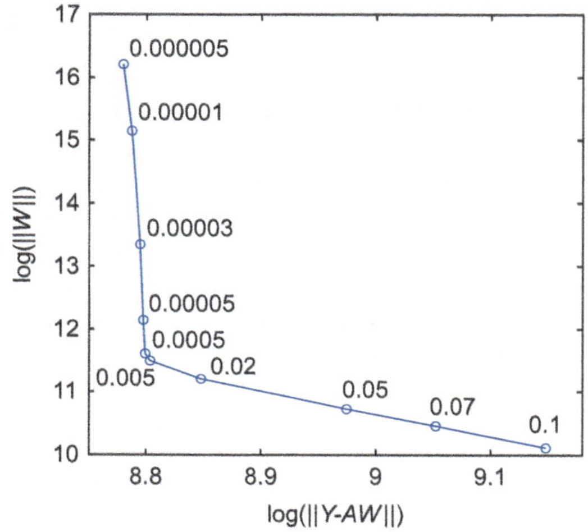

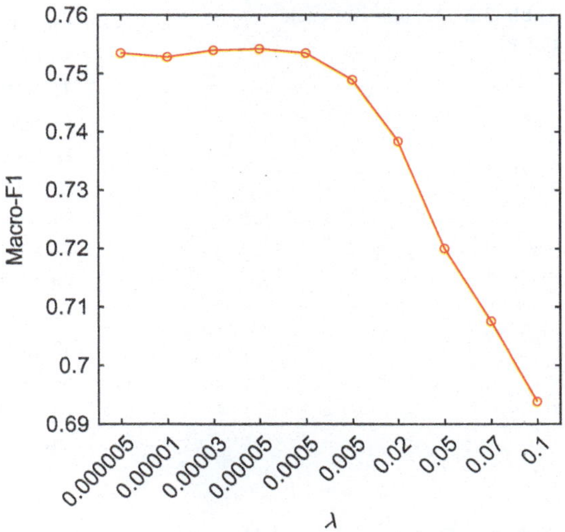

Fig. 3.5 Macro-F1 of different λ for CBL

regularization and over regularization. When λ is greater than 0.005, the model is at over regularization state. When λ is less than 0.0005 , the model is at under regularization state. Thus, we select the parameter λ near the inflection point, i.e., $\lambda = 0.0005$. Macro-$F1$ under different λ on the test set is shown in Fig. 3.5.

From Fig. 3.5, it is found that Macro-$F1$ increases slightly when λ increases from 0.000005 to 0.00005, and reaches peak when $\lambda = 0.00005$. When λ increases from 0.0005 to 0.1, Macro-$F1$ decreases significantly. It shows that over-regularization will not fit to train data properly and the estimation error will be large. On the other hand, under-regularization will fit the train set well but the generalization ability of model is poor. Macro-$F1$ of $\lambda = 0.0005$ selected from the validation set is very close to the peak. The main reason is that the distribution between the validation set and test set are slightly different.

The hyperparameters of BL and CBL on each data set are show in Table 3.1.

Table 3.1 Related parameters for BL and CBL

Model	Parameter	SemEval	Usual	Virus
BL	(n, k)	(41,50)	(22,50)	(18,50)
	(m, q)	(41,50)	(19,50)	(15,50)
	λ	0.0005	0.0003	0.0003
	Activation function	tanh	tanh	tanh
CBL	(n, k)	(41,50)	(22,50)	(18,50)
	(m, q)	(41,50)	(19,50)	(15,50)
	λ	0.0005	0.0005	0.0001
	Activation function	tanh	tanh	tanh

Table 3.2 Accuracy of each model

Model	SemEval	Usual	Virus
SVM	82.52	72.43	70.42
TextCNN	89.29	76.16	77.53
RCNN	88.76	74.90	76.03
Bi-LSTM	87.12	76.24	77.40
Bi-LSTM-ATTN	87.70	76.83	78.06
ERNIE-CNN	–	76.77	77.65
ERNIE-RNN	–	77.37	77.15
DLCF-DCA	88.07	76.41	78.66
Pos-wt	86.98	75.85	75.08
YUN-HPCC	89.31	–	–
SCIA	89.71	–	–
BL	89.53	77.79	77.88
CBL	**90.14**	**78.54**	**79.03**

3.3.4 Experimental Result

The accuracy of baseline methods and our proposed method on three kinds of data sets are shown in Table 3.2, respectively.

From Table 3.2, it is observed that CBL outperforms the baseline methods in accuracy. CBL outperforms SCIA by 0.43 on SemEval, outperforms Bi-LSTM-ATTN by 0.75 on Usual, and outperforms DLCF-DCA by 0.37. The main reason is that CBL can effectively improve the performance by adopting cascading manner for BL.

In addition, it can be found that CBL can improve the performance of TCE more effectively than that of BL on three kinds of data sets. The main reason is that multi-layer mapping of cascading structure can extract more useful information from texts.

Macro-$F1$ of baseline methods and our proposed method on three kinds of data sets are shown in Table 3.3.

In Table 3.3, the experimental results on macro-$F1$ are provided on three data sets. It is seen that SCIA can achieve the best overall performance of 75.66 on SemEval, BL can achieve the best overall performance of 75.25 on Usual, and ERNIE-CNN can achieve the best overall performance of 64.56 on Virus, among the baseline methods. By comparison, the performance of our proposed CBL outperforms SCIA, BL, and ERNIE-CNN by 0.22, 0.54, and 0.63, respectively.

In general, CBL performs better than the other methods in most of cases. The advantages of CBL is summarized in four aspects: (i) Cascading feature nodes and enhancement nodes can help to further extract the deep semantic information of text; (ii) MPNet can provide cascading BL to capture semantic feature and contextual feature; (iii) Solving pseudo inverse matrix by using ridge regression not only can prevent over-fitting, but also can reduce the consumption of computing source; and (iv) MPNet can better reduce distance between sentence embeddings with similar semantics and emotion.

Table 3.3 Macro-$F1$ of each model

Model	SemEval	Usual	Virus
SVM	57.39	66.13	55.49
TextCNN	73.51	72.85	63.29
RCNN	68.03	70.86	57.57
Bi-LSTM	68.34	73.11	60.66
Bi-LSTM-ATTN	69.45	73.94	61.95
ERNIE-CNN	–	72.90	64.56
ERNIE-RNN	–	73.42	63.60
DLCF-DCA	68.22	73.37	64.20
Pos-wt	65.27	72.33	59.54
YUN-HPCC	75.16	–	–
SCIA	75.66	–	–
BL	74.80	75.25	64.41
CBL	**75.78**	**75.79**	**65.19**

Table 3.4 Training time (s) of each method on three data sets

Model	Usual	Virus	SemEval
TextCNN	2012.27	946.99	2370.25
RCNN	2440.22	1031.54	2547.31
Bi-LSTM	4481.99	978.03	5013.67
Bi-LSTM-ATTN	6873.24	1728.66	6758.3
BL	199.58	65.48	226.44
CBL	**193.23**	**63.3**	**219.73**

Table 3.5 Inference time (s) of each method on three data sets

Model	Usual	Virus	SemEval
TextCNN	279.31	94.50	305.33
RCNN	292.84	103.29	321.18
Bi-LSTM	327.02	115.42	338.73
Bi-LSTM-ATTN	314.89	119.46	323.25
DLCF-DCA	369.77	135.51	377.63
Pos-wt	250.39	82.04	283.42
BL	192.63	63.38	217.54
CBL	**188.74**	**61.02**	**210.13**

We also observe that CBL can improve the performance of TEC more effectively than that of BL on three kinds of data sets. This indicates that the multi-layer mapping of cascading structure can better extract the useful information for TEC task.

In addition, we compare the training time and inference time on the test set among baseline methods and CBL. The training time includes the time of generating word embeddings and training the downstream model, and the inference time include the time of generating word embedding and inference process. The species are listed in Tables 3.4 and 3.5, respectively.

From Table 3.4, it is found that the training time of CBL is less than that of the baseline methods. The main reason is that DL-based methods require multiple rounds of training to optimize network parameters. However, cascading BL only needs to optimize the weights of the output layer through ridge regression. From Table 3.5, it can be seen that the inference time of CBL is less than that of the baseline methods. In addition, the training time of CBL is shorter than that of BL. The above results indicate that CBL can not only improve performance but also reduce computational complexity.

3.3.5 Effect of Feature and Enhancement Nodes

To verify the effectiveness of the cascading structure, we discuss it from the following two aspects: (i) assume the number of groups of feature node is the same, but the number of groups of enhancement node is different; and (ii) assume the number of groups of enhancement node is the same, but the number groups of feature node is different. That is, (i) when n=22, m changes from 1 to 28; (ii) when m=19, n changes from 1 to 28. In addition, the number of nodes in each group is set to 50. The relevant experiments are conducted on SemEval and the experimental results are shown in Figs. 3.6 and 3.7.

From Fig. 3.6, it is known that Macro-$F1$ of CBL is higher than that of BL. It illustrates that the multi-layer linear mapping of input by cascading enhancement nodes can effectively improve the performance. With the increase of m, the growth rate of Marco-$F1$ of CBL is larger than that of BL. When m=19, CBL can achieve the best performance.

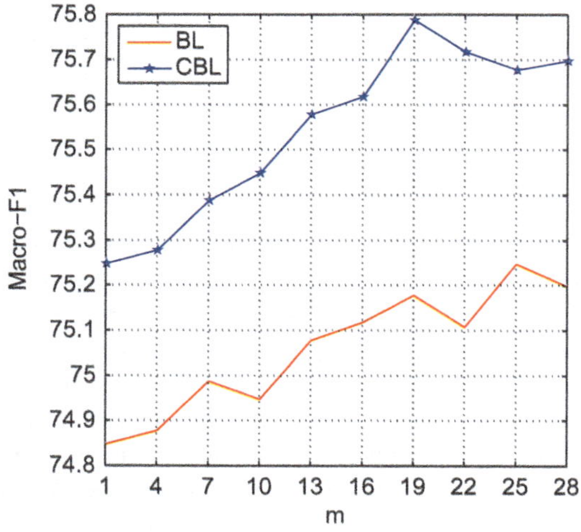

Fig. 3.6 The effect on CBL and BL under different number of groups of enhancement node

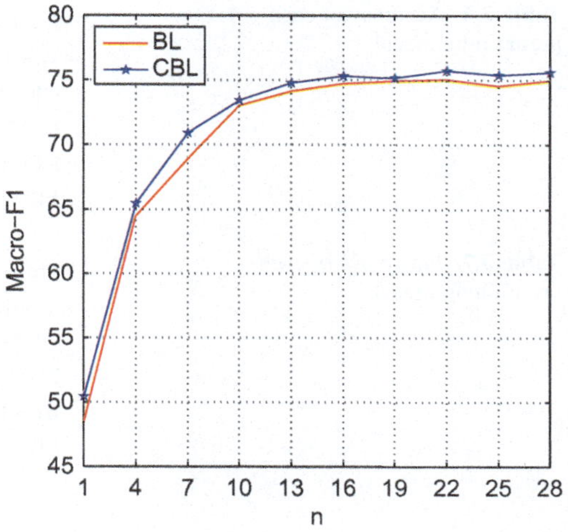

Fig. 3.7 The effect on CBL and BL under different number of groups of feature node

From Fig. 3.7, it is observed that when n=1, CBL with cascading feature nodes is not much improved by comparison with BL. The reason may be that a small number of feature nodes is difficult to capture more useful information. However, it is observed that when n increases from 2 to 13, Macro-$F1$ of CBL and BL increases very quickly and CBL is gradually higher than BL, which shows the efficiency of cascading feature nodes. When n=22, CBL can achieve the best performance.

Thus, according to the above analysis, it can be concluded that when m=19 and n=22, CBL can achieve the best performance on SemEval.

3.3.6 Effect of Pre-training Model

The embedding representation of text is an important factor for the performance of TEC. Selecting a suitable pre-training model can help to obtain better sentence vectors. To analyze the effect on the performance of CBL under different pre-training models,we adopt them as the sentence embeddings of CBL as follows:

- **Word2Vec** [10]: It maps the sparse one-hot word embedding into the dense word embedding by using three-layer network. This method includes CBOW and Skip-Gram. CBOW uses the context to predict the current word, and Skip-Gram uses the current word to predict the context. In our experiment, all of the word embeddings with dimension of 200 are averaged as the sentence embedding.
- **GloVe** [12]: It is a pre-training language model based on word frequency statistics, which uses global statistical information and local statistical information of words to generate word embedding. In the experiment, all of the word embeddings with dimension of 300 are averaged as the sentence embedding.

Table 3.6 Accuracy of each pre-training model

Pre-training model	SemEval	Usual	Virus
Word2Vec	73.19	63.49	67.91
GloVe	80.63	71.34	72.67
ELMo	84.52	72.76	73.47
BERT	88.39	77.52	76.84
MPNet	**90.14**	**78.54**	**79.03**

Table 3.7 Macro-$F1$ of each pre-training model

Pre-training model	SemEval	Usual	Virus
Word2Vec	59.25	53.41	48.34
GloVe	63.03	65.79	53.16
ELMo	66.72	68.89	54.20
BERT	73.31	75.06	64.53
MPNet	**74.67**	**75.79**	**65.19**

- **ELMo** [13]: It generates dynamic word vectors by using multi-layer Bi-LSTM, which solves the polysemous problem in traditional static word embedding. In the experiment, all of the word embeddings with dimensions of 1024 are averaged as the sentence embedding.

The accuracy and Macro-$F1$ of each pre-training model are shown in Tables 3.6 and 3.7, respectively.

From Tables 3.6 and 3.7, it is known that MPNet can achieve the best performance for accuracy and Macro-$F1$. MPNet can outperform the baseline models, which not only shows MPNet can better capture emotional information of text than that of them, but also shows CBL can further extract deep semantic information. Word2Vec and GloVe perform the worst among of them, because they fail to capture the sensitivity of word order in emotion classification. For example, the static sentence embedding of "I like this film, and I don't hate this film" and "I hate this film, and I don't like this film" are similar, but different word orders may make them express different emotions.

3.4 Summary

In this chapter, we explored the current research status of TEC on the basis of DL and proposed a novel method for it to solve the problems, such as long training time, and difficulty in theoretical analysis of DL methods, by using MPNet and cascading BL. The experimental results show that the combination of MPNet and cascading BL can better improve the performance of TEC than that of the traditional DL-based and BL-based methods.

References

1. Abdul-Mageed, M., Ungar, L.: EmoNet: Fine-grained emotion detection with gated recurrent neural networks. In: Proceedings of the 55th Annual Meeting of the Association for Computational Linguistics (Volume 1: Long Papers). pp. 718–728. Vancouver, Canada (Jul 2017)
2. Ahmad, Z., Jindal, R., Ekbal, A., Bhattachharyya, P.: Borrow from rich cousin: transfer learning for emotion detection using cross lingual embedding. Expert Systems with Applications **139**, 112851 (2020)
3. Cao, L., Peng, S., Yin, P., Zhou, Y., Yang, A., Li, X.: A survey of emotion analysis in text based on deep learning. In: 2020 IEEE 8th International Conference on Smart City and Informatization (iSCI). pp. 81–88 (2020)
4. Hansen, P.C.: Analysis of discrete ill-posed problems by means of the l-curve. SIAM review **34**(4), 561–580 (1992)
5. He, H., Xia, R.: Joint binary neural network for multi-label learning with applications to emotion classification. In: Zhang, M., Ng, V., Zhao, D., Li, S., Zan, H. (eds.) Natural Language Processing and Chinese Computing. pp. 250–259. Cham (2018)
6. Ishiwatari, T., Yasuda, Y., Miyazaki, T., Goto, J.: Relation-aware graph attention networks with relational position encodings for emotion recognition in conversations. In: Proceedings of the 2020 Conference on Empirical Methods in Natural Language Processing (EMNLP). pp. 7360–7370 (Nov 2020)
7. Kulshreshtha, D., Goel, P., Kumar Singh, A.: How emotional are you? neural architectures for emotion intensity prediction in microblogs. In: Proceedings of the 27th International Conference on Computational Linguistics. pp. 2914–2926. Santa Fe, New Mexico, USA (Aug 2018)
8. Li, P., Li, J., Sun, F., Wang, P.: Short text emotion analysis based on recurrent neural network. In: Proceedings of the 6th International Conference on Information Engineering (ICIE 2017). pp. 1–5. Association for Computing Machinery, Dalian Liaoning, China (Aug 2017)
9. Ma, F., Zhang, C., Song, D.: Exploiting position bias for robust aspect sentiment classification. In: Findings of the Association for Computational Linguistics: ACL-IJCNLP 2021. pp. 1352–1358 (2021)
10. Mikolov, T., Chen, K., Corrado, G., Dean, J.: Efficient estimation of word representations in vector space. In: Proceedings of the International Conference on Learning Representations (ICLR 2013). pp. 1–12. Arizona, USA (2013)
11. Mohammadi, E., Amini, H., Kosseim, L.: Neural feature extraction for contextual emotion detection. In: Proceedings of the International Conference on Recent Advances in Natural Language Processing (RANLP 2019). pp. 785–794. Varna, Bulgaria (Sep 2019)
12. Pennington, J., Socher, R., Manning, C.: GloVe: Global vectors for word representation. In: Proceedings of the 2014 Conference on Empirical Methods in Natural Language Processing (EMNLP). pp. 1532–1543. Doha, Qatar (Oct 2014)
13. Peters, M.E., Neumann, M., Iyyer, M., Gardner, M., Clark, C., Lee, K., Zettlemoyer, L.: Deep contextualized word representations. In: Proceedings of the 2018 Conference of the North American Chapter of the Association for Computational Linguistics: Human Language Technologies, Volume 1 (Long Papers). pp. 2227–2237. New Orleans, Louisiana (Jun 2018)
14. Reimers, N., Gurevych, I.: Sentence-BERT: Sentence embeddings using Siamese BERT-networks. In: Proceedings of the 2019 Conference on Empirical Methods in Natural Language Processing and the 9th International Joint Conference on Natural Language Processing (EMNLP-IJCNLP). pp. 3982–3992. Hong Kong, China (Nov 2019)
15. Rohanian, O., Taslimipoor, S., Evans, R., Mitkov, R.: WLV at SemEval-2018 task 3: Dissecting tweets in search of irony. In: Proceedings of The 12th International Workshop on Semantic Evaluation. pp. 553–559. New Orleans, Louisiana (Jun 2018)
16. Tafreshi, S., Diab, M.: Emotion detection and classification in a multigenre corpus with joint multi-task deep learning. In: Proceedings of the 27th International Conference on Computational Linguistics (COLING 2018). pp. 2905–2913. Santa Fe, New Mexico, USA (Aug 2018)

17. Xu, M., Zeng, B., Yang, H., Chi, J., Chen, J., Liu, H.: Combining dynamic local context focus and dependency cluster attention for aspect-level sentiment classification. Neurocomputing **478**, 49–69 (2022)
18. Yang, Y., Zhou, D., He, Y., Zhang, M.: Interpretable relevant emotion ranking with event-driven attention. In: Proceedings of the 2019 Conference on Empirical Methods in Natural Language Processing and the 9th International Joint Conference on Natural Language Processing (EMNLP-IJCNLP). pp. 177–187. Hong Kong, China (Nov 2019)
19. Yu, J., Marujo, L., Jiang, J., Karuturi, P., Brendel, W.: Improving multi-label emotion classification via sentiment classification with dual attention transfer network. In: Proceedings of the 2018 Conference on Empirical Methods in Natural Language Processing. pp. 1097–1102. Brussels, Belgium (Oct-Nov 2018)
20. Zhang, Y., Fu, J., She, D., Zhang, Y., Wang, S., Yang, J.: Text emotion distribution learning via multi-task convolutional neural network. In: Proceedings of the 27th International Joint Conference on Artificial Intelligence (IJCAI 2018). pp. 4595–4601 (2018)
21. Zhou, D., Wu, S., Wang, Q., Xie, J., Tu, Z., Li, M.: Emotion classification by jointly learning to lexiconize and classify. In: Proceedings of the 28th International Conference on Computational Linguistics. pp. 3235–3245. Barcelona, Spain (Dec 2020)

Chapter 4
Dual Broad Learning for Textual Emotion Classification

Abstract Although BBL and CBL were proposed in Chaps. 2 and 3, respectively, and they have achieved good results, there is room for improvement of performance of TEC task. Thus, in this chapter, we propose a new method for TCE task, named DBL, which is based on dual BL. DBL consists of two components like BL-BERT and BL-MPNet. As to BL-BERT, texts are input into BERT to generate sentence embedding containing useful information and then the output is input into BL. As to BL-MPNet, texts are input into MPNet to generate sentence embedding containing useful information, and then the output is also input into BL. Finally, it needs to conduct the concatenation for fusing the features from BERT and MPNet.

4.1 Introduction

With the wide application of DL, some researchers adopt dual DL models for some domains such as load forecasting [2], vision processing [8], multi-modal sleep scoring [12], and so on. However, most existing methods for TEC task use the DL models like CNN, attention, GCN, and LSTM, and few of them consider to use dual DL models like dual CNN, dual attention, dual GCN, and dual LSTM.

There are some methods that adopt dual DL models (including dual CNN, dual attention, dual GCN, and dual LSTM) for solving some actual problems. On the basis of dual CNN, Pan et al. [8] adopted a dual convolutional neural network (DualCNN) for low-level vision problems (e.g., super-resolution, edge-preserving filtering, deraining, and dehazing). DualCNN consists of two components, one is a shallow sub-network for estimating structure, and the other is a deep sub-network for estimating details. Zhang et al. [12] adopted a dualCNN to simultaneously process two types of inputs (i.e., temporal or time-frequency) in multi-modal sleep scoring.

On the basis of dual attention, Ding et al. [3] designed the dual attention vision transformers (DaViT), which is a simple yet effective vision transformer architecture. It can capture the global context while maintaining computational efficiency. Xu et al. [11] utilized a dual attention network to generate fine-grained

S. Peng, L. Cao, *Textual Emotion Classification Using Deep Broad Learning*, Socio-Affective Computing 11, https://doi.org/10.1007/978-3-031-67718-2_4

tailored representations for few-shot recognition. It consists of two dual branches including hard- and soft-attentions.

Wang et al. [9] adopted dual attention networks to design a dual attention suppression (DAS) attack, which generates the visually-natural physical adversarial camouflages with strong transferability by suppressing both model and human attention. Zhang et al. [14] proposed a model, called multi-factors aware dual-attentional knowledge tracing (MF-DAKT), which can enrich the question representations and model the learning progress of students based on a dual-attentional mechanism. Nam et al. [7] adopted dual attention networks (DANs) for multimodal reasoning and matching, which can capture the fine-grained interplay between vision and language by jointly leveraging the visual and textual attention mechanisms. DANs can attend to the specific regions in images and words in text through multiple steps and obtain essential information from both modalities.

On the basis of dual GCN, Zhang et al. [13] applied dual GCN to capture the global context by using two orthogonal graphs in a single framework. The first GCN aims to model spatial relationships between pixels in the image, and the second aims to model interdependencies along the channel dimensions of the feature map in network. Wang et al. [10] proposed a model for textbook question answering, called dynamic dual graph networks (DDGNet), which includes diagram graph network (DGN) and textual graph network (TGN). Specifically, DGN aims to avoid ambiguity of visual features for the abstract constituents and help express explicit semantics by combining the text features with positional features of text boxes in the diagram as the node feature. TGN aims to learn contextualized interaction of the useful information in the graph reasoning process by using representation of each sentence in the most related paragraph as the node feature.

Li et al. [5] proposed a dual graph convolutional networks (DualGCN) model for the aspect-based sentiment analysis that considers the complementarity of syntax structures and semantic correlations simultaneously. To alleviate the dependency parsing errors, they designed a SynGCN module with rich syntactic knowledge. To capture the semantic correlations, they designed a SemGCN module with self-attention mechanism. In addition, they adopted the orthogonal and differential regularizer to capture the semantic correlations between words. Fu et al. [4] adopted dual GCN for rain streak removal. Specifically, they designed two graphs to conduct global relational modeling and reasoning. The first GCN aims to explore global spatial relations among pixels in feature maps, and the second GCN aims to model the global relations across the channels.

On the basis of dual LSTM, Chen et al. [1] proposed a dual LSTM-driven semantic understanding method to effectively utilize the remote sensing image scene information. It fused the global features of the image with the features of the target area, and the attention distribution of the target is predicted by using dual LSTM to predict the attention distribution of each geo-object. Ma et al. [6] proposed a time-series data prediction method based on Dual-LSTM. The first LSTM model is established to make a preliminary prediction of the equipment state. The second LSTM model is established for the accumulation of errors in the multi-step prediction process based on the prediction residual data.

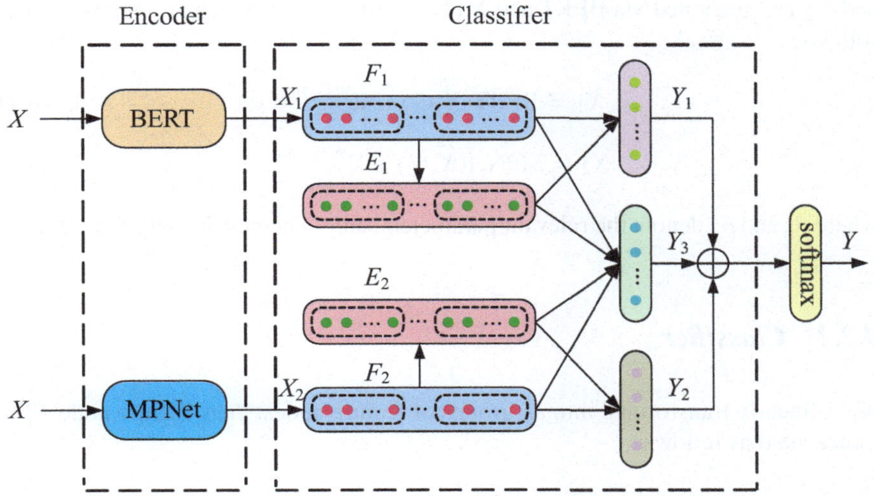

Fig. 4.1 The framework of DBL

Motivated by the better performance of dual DL models in many fields, in this chapter, we will adopt dual BL for TEC task. Specifically, one BL is combined with BERT, the other BL is combined with MPNet. Then, they are integrated to classify the emotion of texts.

4.2 Methodology

DBL is a method for TEC task. It consists of three layers: encoder layer, classifier layer, and prediction layer. The framework of DBL is shown in Fig. 4.1.

In the encoder layer, the pre-training model BERT and MPNet are adopted to obtain the word embeddings of texts. In the classifier layer, dual BLs are utilized to obtain the features of each text. In the prediction layer, the output of BL is input into the softmax to obtain the prediction results.

4.2.1 Encoder

In DBL, we are given a set of training sample data is denoted by $\{X, Y\} \in \mathbb{R}^{N\times(M+C)}$, where N denotes the total number of input samples, M denotes the number of feature dimension, and C denotes the number of emotion classes. X_1

and X_2 are generated via BERT and MPNet, respectively, which are represented as follows:

$$X_1 = \mathrm{BERT}(X,\theta_1) \in \mathbb{R}^{Nl\times 768}, \tag{4.1}$$

$$X_2 = \mathrm{MPNet}(X,\theta_2) \in \mathbb{R}^{Nl\times 768}, \tag{4.2}$$

where θ_1 and θ_2 denote the relevant parameters, and l denotes the length of text.

4.2.2 Classifier

X_1 is linearly transformed into n_1 groups of feature nodes, then, feature node F_1^i is represented as follows:

$$F_1^i = g_1\left(X_1 W_1^{fi} + \beta_1^{fi}\right) \in \mathbb{R}^{N\times v_1}, i = 1, 2, \ldots, n_1, \tag{4.3}$$

where g_1 denotes a linear activation function, v_1 denotes the number of feature nodes in each group, and W_1^{fi} and β_1^{fi} denote the weight matrix and bias matrix of the i-th group of feature nodes generated randomly, respectively.

It is assumed that $F_1 \triangleq \left[F_1^1, F_1^2, \ldots, F_1^{n_1}\right]$ is concatenated by n_1 groups of features nodes, which denotes the linear feature matrix generated by BERT. F_1 are nonlinearly transformed into m_1 groups of enhancement nodes, and then the j-th group of enhancement nodes E_1^j is represented as follows:

$$E_1^j = \varphi_1\left(F_1 W_1^{ej} + \beta_1^{ej}\right) \in \mathbb{R}^{N\times u_1}, j = 1, 2, \ldots, m_1, \tag{4.4}$$

where φ_1 denotes a nonlinear activation function F_1, u_1 denotes the number of enhancement nodes in each group, and W_1^{ej} and β_1^{ej} denote the weight matrix and bias matrix of the j-th group of feature nodes generated randomly, respectively.

It is assumed that $E_1 \triangleq \left[E_1^1, E_1^2, \ldots, E_1^{m_1}\right]$ is concatenated by m_1 groups of enhancement nodes, which denotes the nonlinear feature matrix. F_1 and E_1 are concatenated as a new matrix, and then the total feature matrix S_1 of BERT can be represented as follows:

$$S_1 = [F_1, E_1] \in \mathbb{R}^{N\times(n_1 v_1 + m_1 u_1)}. \tag{4.5}$$

Similarly, sentence embedding X_2 is linearly transformed into n_2 groups of feature nodes, then, feature node F_2^i is represented as follows:

$$F_2^i = g_2\left(X_2 W_2^{fi} + \beta_2^{fi}\right) \in \mathbb{R}^{N\times v_2}, i = 1, 2, \ldots, n_2, \tag{4.6}$$

where g_2 denotes a linear activation function, v_2 denotes the number of feature nodes in each group, and W_2^{fi} and β_2^{fi} denote the weight matrix and bias matrix of the i group of feature nodes generated randomly, respectively.

It is assumed that $F_2 \triangleq \left[F_2^1, F_2^2, \ldots, F_2^{n_2}\right]$ is concatenated by n_2 groups of features nodes, which denotes the linear feature matrix generated by BERT. F_2 are nonlinearly transformed into m_2 groups of enhancement nodes, and then the j-th group of enhancement nodes E_2^j is represented as follows:

$$E_2^j = \varphi_2\left(F_2 W_2^{ej} + \beta_2^{ej}\right) \in \mathbb{R}^{N \times u_2}, j = 1, 2, \ldots, m_2, \tag{4.7}$$

where φ_2 denotes a nonlinear activation function F_2, u_2 denotes the number of enhancement nodes in each group, and W_2^{ej} and β_2^{ej} denote the weight matrix and bias matrix of the j group of feature nodes generated randomly, respectively.

It is assumed that $E_2 \triangleq \left[E_2^1, E_2^2, \ldots, E_2^{m_2}\right]$ is concatenated by m_2 groups of enhancement nodes, which denotes the nonlinear feature matrix. F_2 and E_2 are concatenated as a new matrix, and then the total feature matrix S_2 of BERT can be represented as follows:

$$S_2 = [F_2, E_2] \in \mathbb{R}^{N \times (n_2 v_2 + m_2 u_2)}. \tag{4.8}$$

To fuse the features from BERT and MPNet, it needs to concatenate S_1 and S_2, fusion matrix S_3 is represented as follows:

$$S_3 = [S_1, S_2] \in \mathbb{R}^{N \times (n_1 v_1 + m_1 u_1 + n_2 v_2 + m_2 u_2)}. \tag{4.9}$$

Thus, the final estimated output $\hat{Y}_1$, $\hat{Y}_2$, and $\hat{Y}_3$ are represented as follows:

$$\hat{Y}_1 = S_1 W_1 \in \mathbb{R}^{N \times C}, \tag{4.10}$$

$$\hat{Y}_2 = S_2 W_2 \in \mathbb{R}^{N \times C}, \tag{4.11}$$

$$\hat{Y}_3 = S_3 W_3 \in \mathbb{R}^{N \times C}, \tag{4.12}$$

where W_1, W_2, and W_3 denote the weight output layers of BL-BERT, BL-MPNet,and fusion, respectively.

4.2.3 Prediction

To obtain the optimal weight of output layer, the ridge regression is adopted as the objective optimization function, which is represented as follows:

$$\underset{W_1}{\operatorname{argmin}} \left(\left\| Y_1 - \hat{Y}_1 \right\|_2^2 + \lambda_1 \left\| W_1 \right\|_2^2 \right), \tag{4.13}$$

$$\underset{W_2}{\operatorname{argmin}} \left(\left\| Y_2 - \hat{Y}_2 \right\|_2^2 + \lambda_2 \left\| W_2 \right\|_2^2 \right), \tag{4.14}$$

$$\underset{W_3}{\operatorname{argmin}} \left(\left\| Y_3 - \hat{Y}_3 \right\|_2^2 + \lambda_3 \left\| W_3 \right\|_2^2 \right), \tag{4.15}$$

where λ_1, λ_2, and λ_3 denote the regularization parameters of the output layer from BL-BERT, BL-MPNet, and fusion, respectively.

According to the regularized least square method, W_1, W_2, and W_3 can be represented as follows:

$$W_1 = \left(\lambda_1 I + S_1 {S_1}^{\mathrm{T}} \right)^{-1} {S_1}^{\mathrm{T}} Y_1, \tag{4.16}$$

$$W_2 = \left(\lambda_2 I + S_2 {S_2}^{\mathrm{T}} \right)^{-1} {S_2}^{\mathrm{T}} Y_2, \tag{4.17}$$

$$W_3 = \left(\lambda_3 I + S_3 {S_3}^{\mathrm{T}} \right)^{-1} {S_3}^{\mathrm{T}} Y_3. \tag{4.18}$$

Thus, the output Y of DBL can be obtained by adding the above outputs, which is represented as follows:

$$Y = \operatorname{softmax}(Y_1 + Y_2 + Y_3) \in \mathbb{R}^{N \times C}. \tag{4.19}$$

According to the above description, the entire process of DBL is listed in Algorithm 4.1.

4.3 Experiment and Analysis

To verify the effectiveness of DBL, We conduct comparative experiments on data sets like SMP-2020 EWECT and NLPCC2014 Task 1. We compare the performance of our proposed method with the baseline methods under the same evaluation metrics. The experimental device is a laptop with Intel-i5-9400 2.9 GHz CPU and 16 GB memory, and server with Nvidia V100 GPU.

Algorithm 4.1 The entire process of DBL

Input: X
Output: Output weights Y
$X_1 = \text{BERT}(X,\theta_1) \in \mathbb{R}^{Nl\times768}$
$X_2 = \text{BERT}(X,\theta_2) \in \mathbb{R}^{Nl\times768}$
for $i = 1; i \leq n_1$ **do**
  Compute $F_1^i = g_1\left(X_1 W_1^{fi} + \beta_1^{fi}\right)$
end for
Compute $F_1 \triangleq [F_1^1, F_1^2, \ldots, F_1^{n_1}]$
for $j = 1; i \leq m_1$ **do**
  Compute $E_1^j = \varphi_1\left(F_1 W_1^{ej} + \beta_1^{ej}\right)$
end for
Compute $E_1 \triangleq [E_1^1, E_1^2, \ldots, E_1^{m_1}]$, $S_1 = [F_1, E_1]$
for $i = 1; i \leq n_2$ **do**
  Compute $F_2^i = g_2\left(X_2 W_2^{fi} + \beta_2^{fi}\right)$
end for
Compute $F_2 \triangleq [F_2^1, F_2^2, \ldots, F_2^{n_2}]$
for $j = 1; j \leq m_2$ **do**
  Compute $E_2^j = \varphi_2\left(F_2 W_2^{ej} + \beta_2^{ej}\right)$
end for
Compute $E_2 \triangleq [E_2^1, E_2^2, \ldots, E_2^{m_2}]$, $S_2 = [F_2, E_2]$, $S_3 = [S_1, S_2]$
Compute $Y_1 = S_1 W_1$, $Y_2 = S_2 W_2$, $Y_3 = S_3 W_3$
Compute $Y = \text{softmax}\,(Y_1 + Y_2 + Y_3)$

Table 4.1 Statistics of the data sets

Data set	Emotion	Train	Validation	Test
Usual	Happy	5379	391	1018
	Sad	4990	346	900
	Angry	8344	586	1508
	Fear	1220	87	210
	Surprise	2086	170	373
	Neutral	5749	420	990
Virus	Happy	4423	923	1540
	Sad	649	165	219
	Angry	1322	314	463
	Fear	555	75	190
	Surprise	197	17	68
	Neutral	1460	476	520

4.3.1 Data Set

SMP2020-EWECT consists of two types data sets usual Weibo and virus Weibo, including six classes of emotions (i.e., happy, angry, sad, fear, surprise, neutral). The statistic of emotion distribution is shown in Table 4.1.

Table 4.2 NLPCC 2014-Task 1 emotion distribution

Class	Training	Validation	Test
Happy	4258	366	1630
Like	2805	229	641
Surprise	820	97	259
Disgust	3130	344	679
Fear	299	45	67
Sadness	2478	267	302
Anger	1899	222	244

NLPCC 2014-Task 1 includes seven classes of emotions (i.e., happy, like, surprise, disgust, fear, sadness, and anger). The statistic of emotion distribution is shown in Table 4.2.

4.3.2 Baseline Method

To verify the effectiveness of our proposed DBL for TEC task, we compare it with the baseline methods: SVM, TextCNN, RCNN, Bi-LSTM, BiLSTM-ATTN, Bi-LSTM-ATTN, ERNIE-CNN, and ERNIE-RNN, which have mentioned in Chaps. 2 and 3.

4.3.3 Parameter Setting

For fair comparison, the word embedding generated by BERT with dimension of 768 is adopted as the input for the baseline methods. During the training process, we set the number of training rounds epoch=30 and introduce the early stop method to ensure that the model can converge in appropriate position, thereby preventing over-fitting. In addition, we set the learning rate 0.001, and adopt the strategy of optimizing learning rate, so that the learning rate can be gradually reduced in the training process to obtain better network parameters. All the models are implemented based on the Tensorflow framework, and their hyperparameter settings are shown in Table 4.3.

As to DBL, there is a significant impact from different numbers of feature nodes and enhancement nodes on model performance. If the number of nodes is too small, it may cause the model to be difficult to fully learn; If the number of nodes is too large, it may lead to over-fitting and performance decrease. The function of regularization parameters is to control the optimal solution of the model, and appropriate setting of regularization parameters can effectively prevent over-fitting of the model. Thus, the appropriate hyperparameters are selected based on the performance of the model on the validation set, and the specific parameter settings for DBL are shown in Table 4.4.

Table 4.3 Hyperparameter settings for the training of relevant DL models

Model	Parameter	Value
TextCNN	Kernel size	3/4/5
	Activation function	ReLU
RCNN	Bi-LSTM layer	1
	Bi-LSTM unit	400
	Kernel size	2
	Activation function	ReLU
Bi-LSTM	Bi-LSTM layer	2
	Bi-LSTM unit	400
Bi-LSTM-ATTN	Bi-LSTM layer	1
	Bi-LSTM unit	400
	attention layer	1

Table 4.4 DBL hyperparameter

Parameter	Value
BL-BERT feature node (n_1, q_1)	(20, 50)
BL-BERT enhancement node (m_1, r_1)	(8, 50)
BL-MPNet feature node (n_2, q_2)	(20, 50)
BL-MPNet enhancement node (m_2, r_2)	(8, 50)
Regularization parameter $\lambda_1, \lambda_2, \lambda_3$	0.0001
Nonlinear mapping function	tanh

Table 4.5 Accuracy and macro-$F1$ of each method on usual

Model	Accuracy	Macro-$F1$
SVM	72.43	66.13
TextCNN	76.16	72.85
RCNN	74.90	70.86
Bi-LSTM	76.24	73.11
Bi-LSTM-ATTN	76.83	73.94
ERNIE-CNN	76.77	72.90
ERNIE-RNN	77.37	73.42
BBL	76.51	72.54
CBL	78.54	75.79
DBL	**79.09**	**75.98**

4.3.4 Result and Analysis

Similar to the Chaps. 2 and 3, we adopt the accuracy and macro-$F1$ as the performance evaluation metrics in our experiments. The accuracy and macro-$F1$ of baseline methods and our proposed method on three kinds of data sets are shown in Tables 4.5, 4.6, and 4.7, respectively.

In Table 4.5, the experimental results are provided on Usual. It is known that ERNIE-RNN achieve the best overall performance of 77.37 on accuracy and Bi-LSTM-ATTN achieve the best overall performance of 73.94 on macro-$F1$ among the baseline methods. By comparison, the performance of our proposed DBL

Table 4.6 Accuracy and macro-$F1$ of each method on virus

Model	Accuracy	Macro-$F1$
SVM	70.42	55.49
TextCNN	77.53	63.29
RCNN	76.03	57.57
Bi-LSTM	77.40	60.66
Bi-LSTM-ATTN	78.06	61.95
ERNIE-CNN	77.65	64.56
ERNIE-RNN	77.15	63.60
BBL	76.38	58.48
CBL	79.03	65.19
DBL	**79.55**	**65.92**

Table 4.7 Accuracy and macro-$F1$ of each method on NLPCC 2014 Task 1

Model	Accuracy	Macro-$F1$
SVM	65.90	50.04
TextCNN	71.93	63.75
RCNN	66.38	54.10
Bi-LSTM	67.79	58.04
Bi-LSTM-ATTN	69.41	61.52
DBL	**73.90**	**66.32**

outperforms ERNIE-RNN by 1.72 on accuracy and outperforms Bi-LSTM-ATTN by 2.04 on macro-$F1$.

In Table 4.6, the experimental results are provided on Virus. It is seen that Bi-LSTM-ATTN achieve the best overall performance of 78.06 on accuracy and ERNIE-CNN achieve the best overall performance of 64.56 on macro-$F1$ among the baseline methods. By comparison, the performance of our proposed DBL outperforms Bi-LSTM-ATTN by 1.49 on accuracy and outperforms ERNIE-CNN by 1.36 on macro-$F1$.

In Table 4.7, the experimental results are provided on NLPCC 2014 Task 1. It is seen that TextCNN achieve the best overall performance of 71.93 on accuracy and 63.75 on macro-$F1$ among the baseline methods. By comparison, the performance of our proposed DBL outperforms TextCNN by 1.97 on accuracy and by 2.57 on macro-$F1$.

From Tables 4.5 and 4.6, it is observed that DBL outperforms the baseline methods. In addition, it is observed that DBL can improve the performance of TEC more effectively than that of BL and CBL. The main reason is that there are three advantages of DBL like: (i) using pre-training models with different features of MPNet and BERT to encode text, and training two layers of BL separately, which can effectively extract semantic information at different levels; (ii) The fusion and training of two-layer BL features can effectively improve the information capacity and generalization performance of the model; and (iii) Using the ridge regression to solve the output layer weight can avoid the over-fitting of the model and help to obtain better performance.

Table 4.8 Ablation analysis of DBL on usual

Model	Accuracy	Macro-$F1$
BL-BERT	77.36	74.79
BL-MPNET	77.81	75.23
DBL	**79.09**	**75.98**

Table 4.9 Ablation analysis of DBL on virus

Model	Accuracy	Macro-$F1$
BL-BERT	77.18	64.23
BL-MPNET	77.86	64.40
DBL	**79.55**	**65.92**

Table 4.10 Ablation analysis of DBL on NLPCC 2014 task 1

Model	Accuracy	Macro-$F1$
BL-BERT	73.11	65.57
BL-MPNET	73.42	66.07
DBL	**73.90**	**66.32**

4.3.5 Ablation Analysis

To comprehensively address the impact of each pre-training model on DBL, we conduct ablation analysis on Usual, Virus, and NLPCC 2014 by removing them separately, and investigate their contributions on these data sets. The experimental results are shown in Tables 4.8, 4.9, and 4.10, respectively.

From Tables 4.8, 4.9, and 4.10, it is found that both BL-BERT and BL-MPNET are essential to the strong performance of DBL on these data sets. There is a greater impact from the removing of BL-MPNET than that of BL-BERT, which shows that MPNet is more important for TEC than BERT.

4.3.6 Effect of Combination of Pre-training Model

The embedding representation of text is an important factor for the performance of TEC. A suitable pre-training model can obtain better sentence vectors. To verify the effect on the performance of system under different combination of pre-training model, the following combination are adopted as the sentence embeddings of DBL.

From Tables 4.11, 4.12, and 4.13, it can be observed that BERT+MPNET can achieve the best performance for these two evaluation metrics. Macro-$F1$ of BERT+MPNET outperforms the other combinations, which not only shows BERT+MPNET can better capture emotional information of text than that of the other combinations, but also shows DBL can further extract deep semantic information. GloVe+ELMo performs the worst among of them, and the main reason is that it is unable to solve the sensitivity of word order in emotion classification. For example, the static sentence embedding of "I like it, and I don't hate it" and "I

Table 4.11 Performance of each combination of pre-training model on usual

Model	Accuracy	Macro-$F1$
GloVe+ELMo	73.56	68.09
BERT+GloVe	74.12	68.26
MPNET+GloVe	74.77	71.15
BERT+ELMo	75.32	69.51
MPNET+ELMo	76.18	71.40
BERT+MPNET	**79.09**	**75.98**

Table 4.12 Performance of each combination of pre-training model on virus

Model	Accuracy	Macro-$F1$
GloVe+ELMo	74.42	55.97
BERT+GloVe	76.47	58.19
MPNET+GloVe	77.66	61.20
BERT+ELMo	76.59	58.66
MPNET+ELMo	77.04	60.51
BERT+MPNET	**79.55**	**65.92**

Table 4.13 Performance of each combination of pre-training model on NLPCC 2014 task 1

Model	Accuracy	Macro-$F1$
GloVe+ELMo	64.14	56.27
BERT+GloVe	67.58	58.72
MPNET+GloVe	69.80	61.09
BERT+ELMo	69.51	60.41
MPNET+ELMo	70.19	62.83
BERT+MPNET	**73.90**	**66.32**

hate it, and I don't like it" are the same, but different word orders make them express different emotions.

4.4 Summary

In this chapter, we explored the current research status of dual DL models and proposed an approach for TEC task by using dual BL. The experimental results show that the combination of BL-BERT and BL-MPNet can better improve the performance of TEC than that of BBL and CBL proposed in Chaps. 2 and 3, respectively.

References

1. Chen, J., Dai, X., Zhou, X., Sun, G., Deng, M.: Semantic understanding of geo-objects relationship in high resolution remote sensing image driven by dual lstm. National Remote Sensing Bulletin **25**(5), 1085–1094 (2021)

2. Chen, P., Guo, X., Tong, L., Wan, C.: Seq2seq short-term load forecasting based on double attention mechanism. Journal of Jilin University (Information Science Edition) **41**(2), 251–258 (2023)
3. Ding, M., Xiao, B., Codella, N., Luo, P., Wang, J., Yuan, L.: Davit: Dual attention vision transformers. In: Proceedings of the European Conference on Computer Vision. pp. 74–92 (2022)
4. Fu, X., Qi, Q., Zha, Z., Zhu, Y., Ding, X.: Rain streak removal via dual graph convolutional network. In: Proceedings of the AAAI Conference on Artificial Intelligence. pp. 1352–1360 (2021)
5. Li, R., Chen, H., Feng, F., Ma, Z., Wang, X., Hovy, E.: Dual graph convolutional networks for aspect-based sentiment analysis. In: Proceedings of the 59th Annual Meeting of the Association for Computational Linguistics and the 11th International Joint Conference on Natural Language Processing. pp. 6319–6329 (2021)
6. Ma, Y., Li, C., Yin, Z., Li, M., Chai, A., Zhao, Z.: Production line equipment states prediction method and application based on dual-lstm hybrid model. Journal of Chinese Computer Systems **41**(12), 2470–2474 (2020)
7. Nam, H., Ha, J.W., Kim, J.: Dual attention networks for multimodal reasoning and matching. In: 2017 IEEE Conference on Computer Vision and Pattern Recognition (CVPR). pp. 2156–2164 (2017)
8. Pan, J., Sun, D., Zhang, J., Tang, J., Yang, J., Tai, Y.W., Yang, M.H.: Dual convolutional neural networks for low-level vision. International Journal of Computer Vision **130**, 1440–1458 (2022)
9. Wang, J., Liu, A., Yin, Z., Liu, S., Tang, S., Liu, X.: Dual attention suppression attack: Generate adversarial camouflage in physical world. In: Proceeding of the IEEE/CVF Conference on Computer Vision and Pattern Recognition (CVPR). vol. 37, pp. 8561–8570 (2021)
10. Wang, Y., Liu, J., Ma, J., Zeng, H., Zhang, L., Li, J.: Dynamic dual graph networks for textbook question answering. Pattern Recognition **139**, 109441 (2023)
11. Xu, S., Zhang, F., Wei, X., Wang, J.: Dual attention networks for few-shot fine-grained recognition. In: Proceedings of the AAAI Conference on Artificial Intelligence (AAAI 2022). pp. 74–92911–29192 (2022)
12. Zhang, L., Chen, D., Chen, P., Li, W., Li, X.: Dual-cnn based multi-modal sleep scoring with temporal correlation driven fine-tuning. Neurocomputing **420**, 317–328 (2021)
13. Zhang, L., Li, X., Arnab, A., Yang, K., Tong, Y., Torr, P.: Dual graph convolutional network for semantic segmentation. In: Proceedings of the 30th British Machine Vision Conference (2019), https://bmvc2019.org/programme/detailed-programme/
14. Zhang, M., Zhu, X., Zhang, C., Ji, Y., Pan, F., Yin, C.: Multi-factors aware dual-attentional knowledge tracing. In: Proceedings of the 30th ACM International Conference on Information and Knowledge Management, Queensland, Australia. pp. 2588–2597 (2021)

Chapter 5
Single-Source Domain Adaptation for Emotion Classification Using CNN and Broad Learning

Abstract Single-source domain adaptation (SSDA) for emotion classification aims to leverage useful information in a source domain to help predict emotional polarity in a target domain in a unsupervised or semi-supervised manner. Due to the domain discrepancy, an emotional classifier trained on source domain may not work well on target domain. Many researchers have focused on traditional cross-domain sentiment classification (CDSC), which is a coarse-grained emotion classification method. However, the problem of emotion classification for cross-domain is rarely involved. In this chapter, we propose a novel single-source domain adaptation approach for TEC task, by combining the strength of CNN and BL. We first utilize CNN to extract domain-invariant feature (DIF) and domain-specific feature (DSF) simultaneously, so as to train two more efficient classifiers by employing BL. Then, to take advantage of these two classifiers, we design a co-training model to boost together for them. Finally, we conduct comparative experiments on four data sets to verify the effectiveness of the proposed method.

5.1 Introduction

SSDA-based emotion classification (SSDAEC) is an important task in domain adaptation, which attracts more and more attention from both academia and industry. This has attracted some researchers to explore the domain adaptation method for SSDAEC task. In recent years, most of them have focused on CDSC task and have proposed many methods, which can be divided into two types: the first kind of CDSC methods is proposed by extracting DIF [6, 12, 13, 20, 25, 35, 37, 42], and the other is proposed by extracting DIF and DSF [5, 19, 26, 27, 29, 44].

As to the first kind of CDSC methods, He et al. [12] proposed a domain adaptive semi-supervised learning framework for CDSC task. It can incorporate the unlabeled target data for classifier refinement by combining the entropy minimization with the self-ensemble bootstrapping. Ji et al. [13] presented a method for CDSC task on the basis of Bifurcated-LSTM. This method can extract the domain-invariant sentiment features from the source domain to perform sentiment analysis in different

S. Peng, L. Cao, *Textual Emotion Classification Using Deep Broad Learning*,
Socio-Affective Computing 11, https://doi.org/10.1007/978-3-031-67718-2_5

target domains by taking the advantages of the attention-based LSTM classifiers along with the augmented data set and orthogonal constraints.

Zhou et al. [42] presented a sentiment-aware pre-training model for CDSC task. The model can learn DIF from the domain-invariant sentiment knowledge, such as the sentiment lexicons, emoticons, and ratings. Li et al. [20] adopted the selective adversarial learning (SAL) for the end-to-end aspect-based sentiment analysis. SAL can automatically capture their latent relations by aligning the inferred correlation vectors. In addition, SAL can also achieve word-level adaptation by learning dynamically an alignment weight for each word.

Peng and Zhang [25] design a framework by adopting a weighted version of the domain invariant representation learning (WDIRL) for CDSC task. Du et al. [6] utilized the Wasserstein-based transfer network (WTN) for CDSC task. WTN can share domain-invariant information between the source domain and target domain. WTN can obtain rich knowledge and deep-level semantic information of text by using BERT. In addition, the features can be automatically captured by using RNN with attention mechanism, and the feature representations between the source domain and the target domain can be estimated by computing Wasserstein distance.

Xue et al. [37] proposed a learning method, named deep adversarial mutual learning (DAML) by combining the advantages of adversarial learning and mutual learning and introducing the label probers. DAML consists of two groups of feature extractors, domain discriminators, sentiment classifiers, and label probers. It can improve the performance of CDSC by learning sentiment polarity from an unlabeled target domain. Wang et al. [35] proposed a softly associative transfer learning algorithm (SATL) for CDSC task. In SATL, two non-negative matrix tri-factorizations are integrated into a joint optimization framework.

As to the second kind of CDSC methods, Du et al. [5] designed a domain-distinguish pre-training model to distill the domain-specific features in a self-supervised manner by conducting post-training and adversarial training. Sharma et al. [29] proposed an approach for CDSC task by identifying the significant consistent polarity (SCP) word based on the chi-square test and cosine-similarity between the context vectors of words, and further by training a classifier for the target domain.

Qu et al. [27] presented an adversarial category alignment network (ACAN) to enhance category consistency between source and target domain. In ACAN, the discrepancy of two polarity classifiers is added to provide diverse views by locating ambiguous features near the decision boundaries, which can help to classify the target samples correctly into the corresponding category. Peng et al. [26] presented an approach to extract the domain specific and domain invariant representations of target domain examples by using CMD-based regularizer, and then these two representations are combined by using a co-training framework.

Ziser and Reichart [44] designed a pivot-based language model (PBLM) for the CDSC task. It is a representation learning model that integrates the pivot-based with neural network in a structure aware manner. It can process the information in the text with LSTM. Li et al. [19] adopted a hierarchical attention transfer network (HATN) for CDSC task. HATN consists of two hierarchical attention networks, one called

P-net which aims to find the pivots and the other called NP-net which aims to align the non-pivots by using the pivots as a bridge.

Although existing methods have achieved significant progress, the major limitation of the first kind of methods is that they fail to consider the domain-specific features. As to the other kind of methods, although they consider both domain-invariant and domain-specific features, they may fall into the problem of non-convergence and model collapse, due to adoption of generative adversarial networks [9].

In addition, existing methods for domain adaptation fail to explore fine-grained emotion classification [2], which aims to automatically extract the emotional states (i.e., happy, moving, angry, sad, fear, disgusted, and surprise) of users from different domains like hotel and movie reviews. However, existing methods for emotion classification only consider single domain, and ignore SSDAEC task.

To effectively address the above problems, in this chapter, we propose a novel method, called single-source BL (SBL) for SSDA-based emotion classification by extracting DIF and DSF simultaneously. Then, we train a classifier based on DIF with the labeled data from both source and target domain, and train another classifier based on DSF with only a small number of target domain labeled data. Finally, these two classifiers are boosted each other to co-train the target domain unlabeled data.

5.2 Overview of Domain Adaptation

Under the support of massive data, ML models, especially DL models, have been widely applied in the following fields, such as computer vision, NLP, and so on. The ideal application scenario for ML is to have a large number of labeled training instances, and the training data has the same distribution as the test data. However, in practical applications, there are issues such as high cost of data acquisition and annotation, as well as incomplete, unbalanced, and inaccurate annotation of data sets. Meanwhile, in traditional ML models, it is often assumed that the data is independent and identically distributed, but this assumption does not hold true. Due to the differences in data distribution, the models trained by these algorithms often cannot achieve the expected results in similar new fields, which limits the generalization ability of the models.

Transfer learning [23] aims to utilize the similarity between data, tasks, or models to transfer knowledge learned from a specific domain (i.e., source domain, which refers to a domain with a large amount of labeled data available for model training) to other domains (i.e., target domain, which refers to a domain without labeled data or only a small amount of labeled data), thereby improving the performance of the model on tasks in other domains. Especially when a large amount of labeled data cannot be obtained in the target domain for better machine learning models, model pretraining can be carried out in different but relevant source domains with a large amount of labeled data, and then the trained model can be adjusted and applied to the target domain. This is able to solve the difficult problem for obtaining labeled data of the target domain in practical applications.

Domain Adaptation (DA) [24, 33, 36] refers to the ability to transfer models from the source domain to the target domain by learning the differences between the source domain and the target domain. It can reduce the difference in data distribution between the source domain and the target domain, so as to improve the generalization ability and performance of models. It is one of the most representative method in transfer learning. It is expected to address the issue of scarce annotated data in the target domain, so as to avoid the high cost of model training. There are wide application prospects and significant theoretical research value for DA in computer vision, natural language processing, recommendation systems, and other fields.

5.2.1 Notations and Definitions

In this section, we provide some notations and definitions that are utilized in this book.

Definition 5.1 Source Domain D_s

Given a set of data instances $\{x_{s_1}, x_{s_2}, \ldots, x_{s_n}\}$ and their corresponding labels $\{y_{s_1}, y_{s_2}, \ldots, y_{s_n}\}$, combine the data instances and their corresponding labels as the source domain data, denoted by as follows:

$$D_s = \{(x_{s_1}, y_{s_1}), (x_{s_2}, y_{s_2}), \ldots, (x_{s_n}, y_{s_n})\}. \tag{5.1}$$

Definition 5.2 Target Domain D_t

Given a set of data instances $\{x_{t_1}, x_{t_2}, \ldots, x_{t_m}\}$ and their corresponding labels $\{y_{t_1}, y_{t_2}, \ldots, y_{t_m}\}$, combine the data instances and their corresponding labels as the source domain data, denoted by as follows:

$$D_t = \{(x_{t_1}, y_{t_1}), (x_{t_2}, y_{t_2}), \ldots, (x_{t_m}, y_{t_m})\}. \tag{5.2}$$

Definition 5.3 Source Task T_s

Given the feature vector x_s of an instance in the source domain, the probability $P(y_s|x_s)$ of its class label can be predicted based on the class prediction function $f(x_s)$. This process is called the source task, denoted by as follows:

$$T_s = \{y_s, P(y_s|x_s)\}. \tag{5.3}$$

Definition 5.4 Target Task T_t

Given the feature vector x_t of an instance in the target domain, the probability $P(y_t|x_t)$ of its class label can be predicted based on the class prediction function

$f(x_t)$. This process is called the target task, denoted by as follows:

$$T_t = \{y_t, P(y_t|x_t)\}. \tag{5.4}$$

Definition 5.5 Domain Adaptation

DA is a type of source task that is similar to the target task, that is, $T_s = T_t$. But the data distribution in the source and target domains is different. It is a particular case of transfer learning for learning the prediction function $f(x_t)$ in the target domain with the $(x_{s_1}, y_{s_1}), (x_{s_2}, y_{s_2}), \ldots, (x_{s_n}, y_{s_n})$ and $(x_{t_1}, y_{t_1}), (x_{t_2}, y_{t_2}), \ldots, (x_{t_m}, y_{t_m})$.

5.2.2 *Existing Methods*

The research scenarios of DA can be divided into different dimensions. In this book, we will classify DA methods from the following dimensions: whether the labeled data can be obtained or not, the number of participating source and target domains, the composition of cross domain data feature spaces, data distribution between domains, and DL-based. The overall framework of DA methods classification is shown in Fig. 5.1.

1. According to whether there are labeled data in the source and target domains, DA can be divided into supervised DA, semi-supervised DA, and unsupervised DA.

   Supervised DA represents that there are both labeled source and target data available, that is, $D_s = \{x_s^i, y_s^i\}_{i=1}^{N_s}$, and $D_{tl} = \{x_t^i, y_t^i\}_{i=1}^{N_{tl}}$.

   Semi-supervised DA represents that there are labeled data in source domain and some labeled data in target domain available, that is $D_s = \{x_s^i, y_s^i\}_{i=1}^{N_s}$, $D_{tl} = \{x_t^i, y_t^i\}_{i=1}^{N_{tl}}$, and $D_t = \{x_t^i\}_{i=1}^{N_t}$.

   Unsupervised DA represents that are labeled data in source domain and unlabeled data in target domain available, that is, $D_s = \{x_s^i, y_s^i\}_{i=1}^{N_s}$ and $D_t = \{x_t^i\}_{i=1}^{N_t}$.
2. According to the number of source and target domains, DA can be divided into three categories: single-source DA, multi-source DA, and multi-target DA.

   Single-source DA: Refers to knowledge coming from single source domain is transferred to a single target domain. Traditional supervised and semi-supervised DA methods mostly belong to this category.

   Multi-source DA: Refers to labeled data coming from multiple different source domains, where there may be differences in data distribution between the source and target domains, as well as possible differences in data distribution among multiple source domains.

   Multi-target DA: Refers to knowledge coming from the source domain is transferred to multiple unlabeled target domains, and there are differences in data

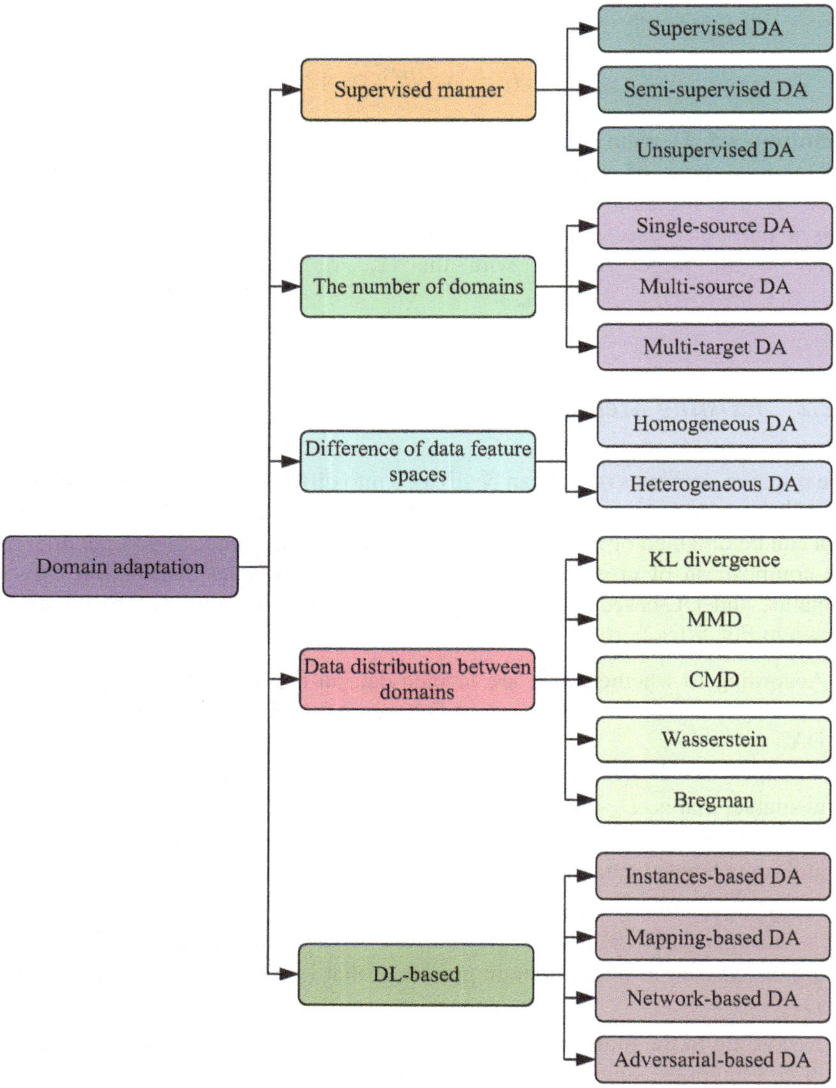

Fig. 5.1 The overall framework of DA methods classification

distribution between the source and target domains, as well as among different target domains. Existing DA methods for TEC almost fail to consider multi-target DA.

3. According to whether the feature spaces between the source and target domain data are the same, DA can be divided into two categories: homogeneous DA and heterogeneous DA.

Homogeneous DA: Refers to the case where there are the same feature space and label space between the source and target domain samples, i.e., $X_s = X_t$, $Y_s = Y_t$, and $d_s = d_t$.

Heterogeneous DA: Refers to the case where there are different feature spaces between the source and target domains. In addition, there are no shared features/labels between the source and target domains, and their dimensions may also be different, i.e., $X_s \neq X_t$, $Y_s \neq Y_t$, and $d_s \neq d_t$.

4. According to the methods for measuring differences in data distribution between domains, DA can be divided into the following classes, such as KL divergence [43], MMD (Maximum Mean Discrepancy) [1, 10, 11], CMD (Central Moment Discrepancy) [38, 39], Wasserstein [17, 22, 30], Bregman [32], and so on. KL divergence is an asymmetric metric of the difference between two probability distributions. Bregman divergence is also an asymmetric distance measurement method. MMD is a widely used metric for measuring differences in DA. CMD can match the high-order center moments of the probability distribution by using the moment difference. Compared to KL divergence, Wasserstein distance considers the basic geometric properties of probability space, and can compare the differences between non-overlapping distributions. Even if the two probability space do not overlap or overlap very little, it can still reflect the distance between two distributions.
5. Since DL algorithms have become popular in various fields, a considerable number of DA methods have been proposed in NLP. Based on the technologies applied in DA, and according to Tan et al in Ref. [34] have classified the deep transfer learning methods into four types, we will divided DL-based DA methods into categories: instances-based DA [4, 15], mapping-based DA [14, 31], network-based DA [40, 41], and adversarial-based DA [7, 18].

Instance-based DA method refers to weighting the samples in the source domain according to the data distribution in the target domain, and then selecting some high confidence samples to add to the training set of target domain.

Mapping-based DA method refers to the assumption that there are distribution differences between source and target domains in the original feature spaces, and then mapping the source and target domains to a new feature space may result in similar distribution.

Network-based DA method refers to pre-training a deep neural network by using a large number of labeled samples from the source domain, then transferring its network structure and connection parameters to the target domain, and fine-tuning the entire or part of neural network by using a small amount of labeled data from the target domain.

Adversarial-based DA method refers to the introduction of adversarial techniques in GAN to find DIF that is suitable for both the source and target domains.

5.3 Methodology

In this section, we first provide the problem definition and introduce maximum mean discrepancy (MMD) [1] metric which can be utilized to measure discrepancy between the probability distribution of any two random variables. Then, we describe our proposed method as to how to effectively obtain DIF and DSF. Finally, we further provide the description on how to fuse these two features by using a co-training model. The framework of our proposed method is shown in Fig. 5.2.

5.3.1 Problem Definition

As to SSDAEC task, we may define two domains D_s and D_t which are given to denote a source domain and a target domain, respectively. Suppose that there is a set of labeled data in the source domain, that is $D_s = \{x_s^i, y_s^i\}_{i=1}^{N_s}$, where N_s denotes the total number of labeled data. In addition, suppose that there is also a set of unlabeled data in the target domain, that is $D_{tu} = \{x_{tu}^i\}_{i=1}^{N_{tu}}$, and some labeled data in the target domain, that is $D_{tl} = \{x_{tl}^i, y_{tl}^i\}_{i=1}^{N_{tl}}$, where N_{tu} denotes the total number of unlabeled data, and N_{tl} denotes the total number of labeled data. The goal of SSDAEC is to

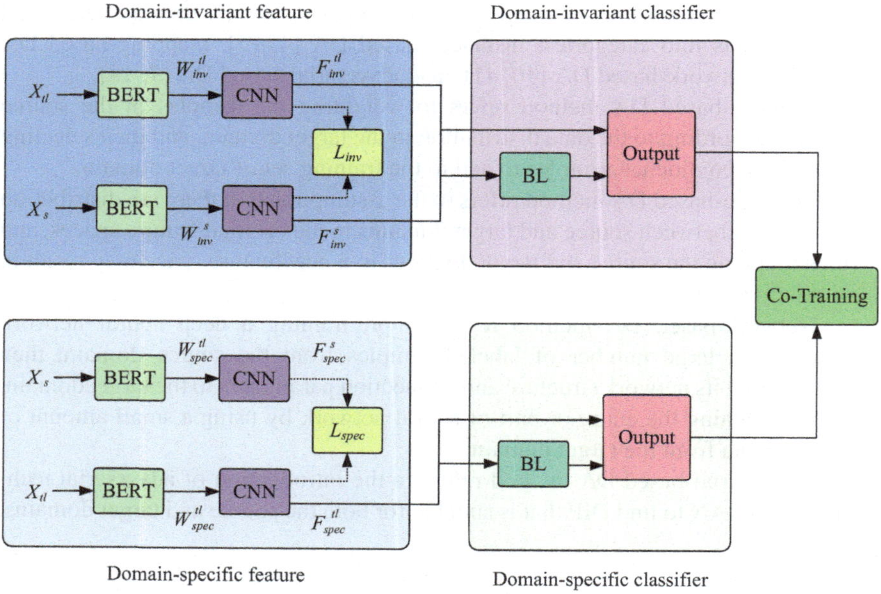

Fig. 5.2 The framework of SBL

train a robust classifier by using the labeled data in the source domain, and then apply it to predict the unlabeled data in the target domain.

5.3.2 *Maximum Mean Discrepancy (MMD)*

MMD was proposed by Borgwardt et al. in Ref. [1] to measure the probability distribution discrepancy between any two random variables. It is differentiable and adopts the Euclidean distance within reproducing kernel Hilbert space (RKHS). MMD is computed between representations of any two domains to reduce the distribution mismatch in the potential space. It is one of the most widely utilized loss function in DA. It is an effective metric to measure discrepancy in the distance of data distribution between source and target domain.

Given two domains D_s and D_t, MMD is defined as follows:

$$\mathrm{MMD}^2(D_s, D_t) = \left\| \frac{1}{N_s} \sum_{i=1}^{N_s} \phi(x_s^i) - \frac{1}{N_t} \sum_{j=1}^{N_t} \phi(x_t^j) \right\|_{\mathcal{H}}^2 , \tag{5.5}$$

where $\phi(\ )$ denotes the nonlinear mapping function that maps the original data to a RKHS, $\mathcal{H}$ denotes a high dimensional RKHS, $\|\phi\|_{\mathcal{H}} \leq 1$ denotes a set of functions in $\mathcal{H}$, and N_t denotes the number of labeled and unlabeled data in target domain.

5.3.3 *Feature Extraction*

To simultaneously extract DIF and DSF, it needs to map the data into a domain-invariant space and a domain-specific space by using two different mappers BERT_{inv} and BERT_{spec}, respectively.

We first use BERT to generate word vectors V_{inv}^s and V_{spec}^s of X_s, which are represented as follows:

$$\begin{aligned} V_{inv}^s &= \mathrm{BERT}_{inv}(X_s; \theta_{inv}^{\mathrm{BERT}}) \in \mathbb{R}^{(N_s l)\times 768}, \\ V_{spec}^s &= \mathrm{BERT}_{spec}(X_s; \theta_{spec}^{\mathrm{BERT}}) \in \mathbb{R}^{(N_s l)\times 768}, \end{aligned} \tag{5.6}$$

where BERT_{inv} and BERT_{spec} denote the representations of DIF and DSF, respectively, which are encoded by BERT; $\theta_{inv}^{\mathrm{BERT}}$ and $\theta_{spec}^{\mathrm{BERT}}$ denote their corresponding parameters, and l denotes the length of text.

Then, we use BERT to generate word vectors V^{tl}_{inv} and V^{tl}_{spec} of X_{tl}, which are represented as follows:

$$V^{tl}_{inv} = \text{BERT}_{inv}(X_{tl}; \theta^{\text{BERT}}_{inv}) \in \mathbb{R}^{(N_{tl}l)\times 768},$$
$$V^{tl}_{spec} = \text{BERT}_{spec}(X_{tl}; \theta^{\text{BERT}}_{spec}) \in \mathbb{R}^{(N_{tl}l)\times 768}. \quad (5.7)$$

Based on these representations, n-gram features are extracted by conducting the convolution and max pooling operations of CNN, respectively. V^{s}_{inv} and V^{s}_{spec} can be transformed to F^{s}_{inv} and F^{s}_{spec}, respectively, which are represented as follows:

$$F^{s}_{inv} = \text{CNN}_{inv}(V^{s}_{inv}; \theta^{\text{CNN}}_{inv}) \in \mathbb{R}^{N_s\times q},$$
$$F^{s}_{spec} = \text{CNN}_{spec}(V^{s}_{spec}; \theta^{\text{CNN}}_{spec}) \in \mathbb{R}^{N_s\times q}, \quad (5.8)$$

where CNN_{inv} and CNN_{spec} denote the representations of DIF and DSF, respectively, which are encoded by using CNN; $\theta^{\text{CNN}}_{inv}$ and $\theta^{\text{CNN}}_{spec}$ denote their corresponding parameters, and q denotes the dimension of convolution output.

Similarly, V^{tl}_{inv} and V^{tl}_{spec} can be transformed to F^{tl}_{inv} and F^{tl}_{spec}, respectively, which are represented as follows:

$$F^{tl}_{inv} = \text{CNN}_{inv}(V^{tl}_{inv}; \theta^{\text{CNN}}_{inv}) \in \mathbb{R}^{N_{tl}\times q},$$
$$F^{tl}_{spec} = \text{CNN}_{spec}(V^{tl}_{spec}; \theta^{\text{CNN}}_{spec}) \in \mathbb{R}^{N_{tl}\times q}. \quad (5.9)$$

As to DIF, we hope it can encode features shared by both source and target domains. From the probability distribution view, we hope that the distributions of the mapped outputs obtained by DIF from source and target data are similar. Thus, we utilize MMD regularizer onto the features of source data F^{s}_{inv} and target data F^{tl}_{inv}. The corresponding loss is defined as follows:

$$L_{sim} = \text{MMD}\left(F^{s}_{inv}, F^{tl}_{inv}\right). \quad (5.10)$$

The distribution discrepancy between F^{s}_{inv} and F^{tl}_{inv} can be reduced by minimizing the loss L_{sim}, which contributes to obtain DIF.

As to DSF, we hope that it extracts features only from target domain. That is, these features should usually emerge in target domain while hardly emerge in source domain. Thus, we utilize MMD regularizer onto the features of source data F^{s}_{spec} and target data F^{tl}_{spec}. The corresponding loss is defined as follows:

$$L_{diff} = -\text{MMD}\left(F^{s}_{spec}, F^{tl}_{spec}\right). \quad (5.11)$$

The distribution discrepancy between F^{s}_{spec} and F^{tl}_{spec} can be enlarged by minimizing loss L_{diff}, which contributes to obtain DSF.

5.3.4 BL-Based Classifier

To further mine the useful information of text, we adopt BL to capture the features for both source and target domains data, and design two classifiers including domain-invariant classifier (DIC) based on DIF and domain-specific classifier (DSC) based on DSF.

As to DIC, we conduct a nonlinear mapping for F_{inv}^{s} and F_{inv}^{tl} into n_{inv} groups of enhanced nodes. Thus, the i-th group of enhanced nodes can be represented as follows:

$$E_{inv}^{i} = \varphi\left(W_{inv}^{i}\left[F_{inv}^{s}, F_{inv}^{tl}\right] + \beta_{inv}^{i}\right), i = 1, 2, \ldots, n_{inv}, \tag{5.12}$$

where φ denotes the nonlinear activation function, and W_{inv}^{i} and β_{inv}^{i} denote the weight matrix and bias matrix generated randomly, respectively.

All the nonlinear features of text data can be denoted by the matrix of n_{inv} groups of enhancement nodes, which is represented as follows:

$$E_{inv} = \left[E_{inv}^{1}, E_{inv}^{2}, \ldots, E_{inv}^{n_{inv}}\right]. \tag{5.13}$$

Thus, the DIC's final estimated output $\hat{Y}_{inv}$ can be represented as follows:

$$\begin{aligned} \hat{Y}_{inv} &= \left[F_{inv}^{s}, F_{inv}^{tl}, E_{inv}\right] W_{inv}^{\mathrm{BL}} \\ &= A_{inv} W_{inv}^{\mathrm{BL}}, \end{aligned} \tag{5.14}$$

where A_{inv} denotes all the input features of DIC, and W_{inv}^{BL} denotes the weight of output layer of DIC.

As DSC only needs to classify target domain data, we conduct nonlinear mapping for F_{spec}^{tl} into n_{spec} groups of enhanced nodes. Thus, the j-th group of enhanced nodes can be represented as follows:

$$E_{spec}^{j} = \varphi\left(W_{spec}^{j} F_{spec}^{tl} + \beta_{spec}^{j}\right), j = 1, 2, \ldots, n_{spec}, \tag{5.15}$$

where W_{spec}^{j} and β_{spec}^{j} denote weight matrix and bias matrix generated randomly, respectively.

All the nonlinear features of text data can be represented by the matrix of n_{spec} groups of enhancement nodes, which is described as follows:

$$E_{spec} = \left[E_{spec}^{1}, E_{spec}^{2}, \ldots, E_{spec}^{n_{spec}}\right]. \tag{5.16}$$

Thus, the DSC's final estimated output $\hat{Y}_{spec}$ can be represented as follows:

$$\begin{aligned}\hat{Y}_{spec} &= \left[F_{spec}^{tl}, E_{spec}\right] W_{spec}^{\mathrm{BL}} \\ &= A_{spec} W_{spec}^{\mathrm{BL}},\end{aligned} \tag{5.17}$$

where A_{spec} denotes all the input features of DSC, and W_{spec}^{BL} denotes the weight of output layer of DSC.

5.3.5 *Co-Training*

The co-training process of this method includes two steps: (i) updating for the corresponding parameters for DIF and DSF; (ii) updating for the classifiers DIC and DSC.

As to DIF, the goal of training process is to minimize the following loss with respect to parameters W_{inv}^{BERT} and W_{inv}^{CNN}:

$$L_{inv} = L_{sim}\left(\theta_{inv}^{\mathrm{BERT}}, \theta_{inv}^{\mathrm{CNN}}\right) + \alpha L_c\left(\theta_{inv}^{\mathrm{BERT}}, \theta_{inv}^{\mathrm{CNN}}\right), \tag{5.18}$$

where α denotes the weight of L_c, $L(\theta)$ denotes that L_{inv} is optimized on the parameter θ during training, and L_c denotes the classification loss on DIF.

L_c is applied to measure the difference between real label and predicted label of the source domain and target domain. Thus, it can be represented as follows:

$$L_c = \frac{1}{N_s + N_{tl}} \sum_{i=1}^{N_s} -y_s^i \log P\left(y_s^i \middle| F_{inv}^{si}\right) + \frac{1}{N_s + N_{tl}} \sum_{j=1}^{N_{tl}} -y_{tl}^j \log P\left(y_{tl}^j \middle| F_{inv}^{tlj}\right), \tag{5.19}$$

where $P()$ denotes the probability function, and F_{inv}^{si} and F_{inv}^{tlj} denote the one-hot encoding of the class label for the i-th source example F_{inv}^{s} and the j-th target example F_{inv}^{tl}, respectively.

As to DSF, the goal of training is to minimize the following loss with respect to parameters $\theta_{spec}^{\mathrm{BERT}}$ and $\theta_{spec}^{\mathrm{CNN}}$:

$$L_{spec} = L_{diff}\left(\theta_{spec}^{\mathrm{BERT}}, \theta_{spec}^{\mathrm{CNN}}\right) + \gamma L_t\left(\theta_{spec}^{\mathrm{BERT}}, \theta_{spec}^{\mathrm{CNN}}\right), \tag{5.20}$$

where γ denotes the weight of loss L_t, which is utilized to measure the real label and predicted label of target domain.

Thus, L_t can be represented as follows:

$$L_t = \frac{1}{N_{tl}} \sum_{i=1}^{N_{tl}} -Y_{tl}^i \log fc\left(y_{tl}^i \middle| F_{spec}^{tli}\right). \tag{5.21}$$

As to DIC, we need to obtain an appropriate W_{inv}^{BL} so that the difference between Y_{inv} and $\hat{Y}_{inv}$ is as small as possible. Thus, the ridge regression is utilized as objective function, which is represented as follows:

$$\underset{W_{inv}^{\mathrm{BL}}}{\operatorname{argmin}} \left(\left\| Y_{inv} - \hat{Y}_{inv} \right\|_2^2 + \lambda_1 \left\| W_{inv}^{\mathrm{BL}} \right\|_2^2 \right), \tag{5.22}$$

where λ_1 denotes regularization parameter and Y_{inv} denotes the actual output of DIC.

As to DSC, similarly, the objective function is represented as follows:

$$\underset{W_{spec}^{\mathrm{BL}}}{\operatorname{argmin}} \left(\left\| Y_{spec} - \hat{Y}_{spec} \right\|_2^2 + \lambda_2 \left\| W_{spec}^{\mathrm{BL}} \right\|_2^2 \right), \tag{5.23}$$

where λ_2 denotes the regularization parameter and Y_{spec} denotes the actual output of DSC.

Thus, the ridge regression is adopted as the objective function to obtain an optimal solution W_{inv}^{BL}, which is represented as follows:

$$W_{inv}^{\mathrm{BL}} = \left(\lambda_1 I + A_{inv} {A_{inv}}^{\mathrm{T}}\right)^{-1} {A_{inv}}^{\mathrm{T}} Y_{inv}. \tag{5.24}$$

Similarly, to obtain an optimal solution W_{spec}^{BL}, it is represented as follows:

$$W_{spec}^{\mathrm{BL}} = \left(\lambda_2 I + A_{spec} {A_{spec}}^{\mathrm{T}}\right)^{-1} {A_{spec}}^{\mathrm{T}} Y_{spec}. \tag{5.25}$$

Here, we regard DIF and DSF as two independent spaces for target domain data. On the basis of DIF, we train DIC with respect to parameter W_{inv}^{BL}. In addition, On the basis of DSF, we train DSC with respect to parameter W_{spec}^{BL}. Because the distribution of source data is compatible with that of the target data in domain-invariant space, we employ both enriched source labels and small amount of target labels to train DIC, and utilize only the target labels to train DSC. The entire training process is described in Algorithm 5.1.

Algorithm 5.1 The training process of domain adaptation

Require: Labeled source domain data X_s, labeled target domain data X_{tl}, unlabeled target domain data X_{tu}

Ensure: W_{inv}^{BL} and W_{spec}^{BL}

```
Repeat
    Obtain DIF based on X_s and X_tl;
    Train DIC based on F^s_inv and F^tl_inv;
    Utilize DIC to predict X_tu;
    Select p samples X^inv_tu with the highest confidence from X_tu;
    Obtain DSF based on X_s and X_tl;
    Train DSC based on F^tl_spec;
    Utilize DSC to predict X_tu;
    Select p samples X^spec_tu with the highest confidence from X_tu;
    Delete samples X^inv_tu ∪ X^spec_tu from X_tu, and add them to X_tl;
until stopping criterion is met
```

5.4 Experiment and Analysis

5.4.1 Data Set and Baseline Method

Since there is a lack of public data set for cross-domain emotion classification, we have collected product reviews as the experimental data set from the well-known Chinese E-commerce platforms, such as Douban,[1] Ctrip,[2] Jingdong,[3] and Taobao.[4] These data sets contain four domains: movie (M), hotel (H), electronics (E), and clothing (C). There is a total number of 41,628 reviews for four data sets, and there are seven emotion classes, including happy, moving, angry, sad, fear, disgusted, and surprise. The detailed statistics are shown in Table 5.1.

According to these data, we constructed twelve cross-domain seven classification tasks. In each domain adaptation task, there are 1000 labeled source domain examples, 1000 unlabeled target domain examples, and 50 labeled target domain examples for training. In addition, to effectively fine-tune the hyper-parameters, we

Table 5.1 Statistics of data set

Name	Happy	Moving	Sad	Angry	Fear	Disgusted	Surprise	Total
Clothing	1699	1035	1819	1483	1014	2014	1237	10301
Electronic	2316	1453	1727	1602	1405	1405	1410	11318
Hotel	1426	1498	1661	1537	1175	1456	1256	10009
Movie	1500	1464	1375	1669	1337	1145	1510	10000

[1] https://www.douban.com/.

[2] https://www.ctrip.com/.

[3] https://www.jd.com/.

[4] https://www.taobao.com/.

randomly select 500 target examples as developing data set, and treat the remainder examples as testing data set. All of the baseline methods and SBL adopt this setting.

We compare our proposed method with the following baselines for SSDAEC task, including Bi-LSTM [28], TextCNN [16], DANN [8], PBLM [44].

- **Bi-LSTM**: It is a non-domain-adaptive method which is trained in source domain and predicts in target domain directly. It is designed on the basis of Bi-LSTM.
- **TextCNN**: It is a convolutional neural network for emotion classification without considering contextual information of text.
- **DANN**: It is a method which exploits a domain classifier to minimize the discrepancy between two domains via an adversarial training manner.
- **PBLM**: It is a representation learning model which exploits the structure of the input text.

5.4.2 Implementation Detail

In our experiments, all the word embeddings from sentences are initialized as 768 dimension vectors via BERT. SBL was implemented with the Chinese BERT pre-training model proposed by Cui et al. [3], which is composed of twelve Transformer blocks and is pre-trained on large number of Chinese corpus (including Chinese Wikipedia, news, Q&A, etc.). The convolution kernel size of CNN is set to two, three, and four, and the dimension of convolution kernel is set to 100. The enhancement nodes of BL consist of 20 groups (i.e., $n_{inv} = n_{spec} = 20$), 50 nodes in each group (i.e., $r_{inv} = r_{spec} = 50$), and tanh activation function.

Model optimization was implemented by using AdamW update strategy [21] with initial learning rate set to 2e-5 and weights decay set to 0.01. The bandwidth and number of the radial basis function (RBF) of MMD are set to 5 and 10, respectively. The hyperparameters corresponding to the best performance on the validation set are obtained by the grid search, and the weight of the loss items set $\alpha = \gamma = 0.5$, the factor for each iteration of co-training is set $p = 5$, and the regularization parameter is set $\lambda_1 = \lambda_2 = 0.001$.

5.4.3 Main Result

To demonstrate the effectiveness of SBL, we compare it with other state-of-the-art methods on SSDAEC task. In addition, to demonstrate the effects of domain adaptation more intuitively, we also compare our proposed SBL with a variant without DSF, denoted by SBL-s. We adopt the classification accuracy to evaluate the models. The experimental results of all methods are shown in Table 5.2, and the overall comparison for accuracy is shown in Fig. 5.3.

Table 5.2 The accuracy of cross-domain emotion classification by different methods

Source→Target	Bi-LSTM	TextCNN	DANN	PBLM	SBL-s	SBL
C→M	45.57	68.11	71.18	73.43	77.14	**80.09**
H→M	63.81	78.29	80.51	80.39	84.11	**85.34**
E→M	51.19	70.30	74.20	77.63	82.54	**83.16**
C→H	58.97	78.80	82.80	85.74	91.22	**92.59**
M→H	61.96	86.92	88.55	87.93	90.61	**92.77**
E→H	55.65	79.31	79.46	81.25	83.85	**86.09**
C→E	60.93	83.12	87.02	88.31	91.28	**92.53**
M→E	51.61	73.02	81.53	82.90	89.34	**92.01**
H→E	60.91	80.56	84.42	83.86	87.21	**90.47**
E→C	59.38	77.31	79.37	80.89	84.50	**84.98**
M→C	53.90	71.76	72.68	72.77	78.90	**80.06**
H→C	55.41	63.01	67.43	71.52	74.02	**76.80**
Average	56.61	75.88	79.10	80.55	84.56	**86.41**

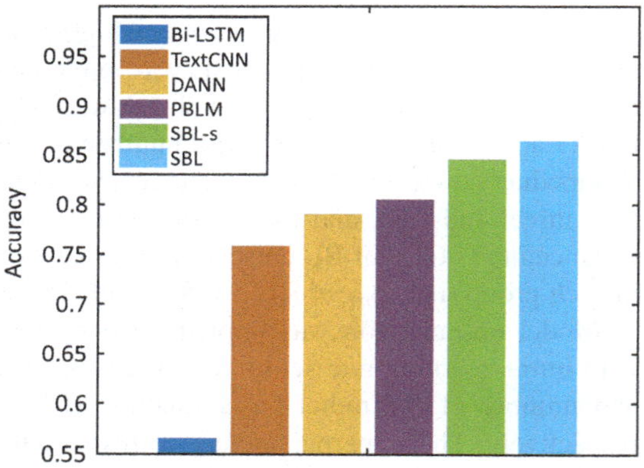

Fig. 5.3 The overall comparison for the accuracy of different methods

Table 5.2 shows the classification accuracy of different methods on four data sets. We evaluate our method over 12 transfer pairs, on a total number of 41,628 testing samples. Our proposed SBL consistently achieves the best performance on most of tasks. Bi-LSTM only achieves 56.61% on average because of its poor extraction ability of domain adaptation features. Compared to TextCNN, DANN and PBLM, SBL outperforms TextCNN by 10.53%, DANN by 7.31%, and PBLM by 5.86% on average, respectively. The possible reasons are that SBL can better exploit automatically domain-invariant and domain-specific representations with BERT, CNN, and BL, which contribute to extract DIF and DSF.

The average accuracy of SBL-s is 84.56%, which is 1.85% lower than SBL. Through the comparison between SBL-s and SBL, we can conclude that SBL can better improve the accuracy of emotion classification.

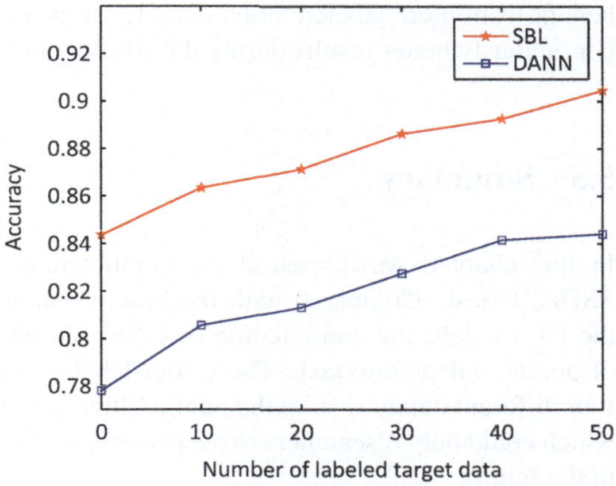

Fig. 5.4 The influence on accuracy under the different number of labeled target data on the task H→E

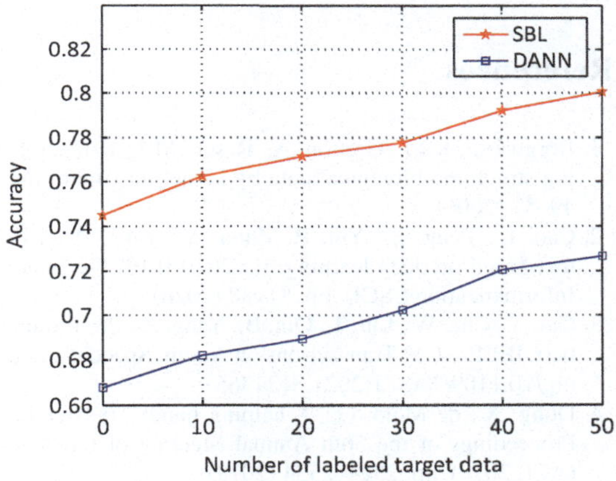

Fig. 5.5 The influence on accuracy under the different number of labeled target data on the task M→C

5.4.4 Effect of the Number of Labeled Target Data

In this subsection, we also provide an analysis under the effect of the number of labeled target data to our proposed SBL. In Figs. 5.4 and 5.5, we show the comparison between DANN and SBL under the data setting that some labeled target data are randomly selected and then mixed them with the training data. Here, we provide the experimental results on two transferring methods, while a similar trend can be obtained in other pairs. The influence on accuracy from the different number of labeled target data on the task H→E and M→C are shown in Figs. 5.4 and 5.5, respectively.

From Figs. 5.4 and 5.5, it is found that when the number of randomly selected labeled target data increases, the difference between two methods almost remains unchanged, and SBL can also keep the advantageous position. These trends show that our proposed SBL is more effective with little labeled target data and can further benefit from more labeled target data. It can be easily observed that SBL can show continuously better results during the whole training process.

5.5 Summary

In this chapter, we discussed the combination advantage of CNN and BL for SSDAEC task. Compared with the baseline methods extracted DIF and DSF via the DL models, the combination of CNN and BL could contribute to the solution of domain adaptation task. These four labeled real-world data sets collected from four different domains from the public Chinese E-commerce platforms are involved, which could help researchers to deeply explore better domain adaptation algorithms in the future.

References

1. Borgwardt, K.M., Gretton, A., Rasch, M.J., Kriegel, H.P., Schlkopf, B., Smola, A.J.: Integrating structured biological data by kernel maximum mean discrepancy. Bioinformatics **22**(14), 49–57 (2006)
2. Cao, L., Peng, S., Yin, P., Zhou, Y., Yang, A., Li, X.: A survey of emotion analysis in text based on deep learning. In: 2020 IEEE 8th International Conference on Smart City and Informatization (iSCI). pp. 81–88 (2020)
3. Cui, Y., Che, W., Liu, T., Qin, B., Yang, Z.: Pre-training with whole word masking for chinese bert. IEEE/ACM Transactions on Audio, Speech, and Language Processing (2021). https://doi.org/10.1109/TASLP.2021.3124365
4. Dong, X., de Melo, G.: A helping hand: Transfer learning for deep sentiment analysis. In: Proceedings of the 56th Annual Meeting of the Association for Computational Linguistics (ACL 2018). pp. 2524–2534 (2018)
5. Du, C., Sun, H., Wang, J., Qi, Q., Liao, J.: Adversarial and domain-aware BERT for cross-domain sentiment analysis. In: Proceedings of the 58th Annual Meeting of the Association for Computational Linguistics. pp. 4019–4028 (2020)
6. Du, Y., He, M., Wang, L., Zhang, H.: Wasserstein based transfer network for cross-domain sentiment classification. Knowledge-Based Systems **204**, 106162 (2020)
7. Duan, X., Zhou, Y., Jing, C., Zhang, L., Chen, R.: Cross-domain sentiment classification based on transfer learning and adversarial network. In: Proceedings of the IEEE 4th International Conference on Computer and Communications. pp. 2302–2306 (2018)
8. Ganin, Y., Ustinova, E., Ajakan, H., Germain, P., Larochelle, H., Laviolette, F., Marchand, M., Lempitsky, V.: Domain-adversarial training of neural networks. The journal of machine learning research **17**(1), 2096–2030 (2016)
9. Goodfellow, I., Pouget-Abadie, J., Mirza, M., Xu, B., Warde-Farley, D., Ozair, S., Courville, A., Bengio, Y.: Generative adversarial nets. In: Proceedings of the 27th International Conference on Neural Information Processing Systems. pp. 2672–2680 (2014)
10. Gretton, A., Borgwardt, K.M., Rasch, M.J., Scholkopf, B., Smola, A.J.: A kernel method for the two-sample-problem. In: Proceedings if the Advances in Neural Information Processing Systems (NIPS 2007). pp. 513–520 (2007)
11. Gretton, A., Borgwardt, K.M., Rasch, M.J., Scholkopf, B., Smola, A.: A kernel two-sample test. Journal of Machine Learning Research **13**, 723–773 (2012)

12. He, R., Lee, W.S., Ng, H.T., Dahlmeier, D.: Adaptive semi-supervised learning for cross-domain sentiment classification. In: Proceedings of the 2018 Conference on Empirical Methods in Natural Language Processing. pp. 3467–3476. Brussels, Belgium (2018)
13. Ji, J., Luo, C., Chen, X., Yu, L., Li, P.: Cross-domain sentiment classification via a bifurcated-lstm. In: Pacific-Asia Conference on Knowledge Discovery and Data Mining. pp. 681–693 (2018)
14. Jia, X., Jin, Y., Li, N., Su, X., Cardiff, B., Bahanu, B.: Words alignment based on association rules for cross-domain sentiment classification. Frontiers of Information Technology & Electronic Engineering **19**(2), 260–272 (2018)
15. Khan, F., Qamar, U., Bashir, S.: Enhanced cross-domain sentiment classification utilizing a multi-source transfer learning approach. Soft Computing **23**, 5431–5442 (2019)
16. Kim, Y.: Convolutional neural networks for sentence classification. In: Proceedings of the 2014 Conference on Empirical Methods in Natural Language Processing (EMNLP). pp. 1746–1751. Doha, Qatar (2014)
17. Lee, C., Batra, T., Baig, M.H., Ulbricht, D.: Sliced wasserstein discrepancy for unsupervised domain adaptation. In: Proceedings of IEEE Conference on Computer Vision and Pattern Recognition (CVPR 2019). pp. 10285–10295 (2019)
18. Li, Z., Li, X., Wei, Y., Bing, L., Zhang, Y., Yang, Q.: Transferable end-to-end aspect-based sentiment analysis with selective adversarial learning. In: Proceedings of the 2019 Conference on Empirical Methods in Natural Language Processing (EMNLP) and the 9th International Joint Conference on Natural Language Processing. pp. 5559–5568 (2019)
19. Li, Z., Wei, Y., Zhang, Y., Yang, Q.: Hierarchical attention transfer network for cross-domain sentiment classification. In: Proceedings of the 32nd AAAI Conference on Artificial Intelligence. pp. 5852–5859 (2018)
20. Li, Z., Li, X., Wei, Y., Bing, L., Zhang, Y., Yang, Q.: Transferable end-to-end aspect-based sentiment analysis with selective adversarial learning. In: Proceedings of the 2019 Conference on Empirical Methods in Natural Language Processing and the 9th International Joint Conference on Natural Language Processing (EMNLP-IJCNLP). pp. 4590–4600. Hong Kong, China (2019)
21. Loshchilov, I., Hutter, F.: Fixing weight decay regularization in adam (2018), https://openreview.net/forum?id=rk6qdGgCZ
22. Luo, Y., Lu, B.: Wasserstein-distance-based multi-source adversarial domain adaptation for emotion recognition and vigilance estimation. In: Proceedings of the IEEE International Conference on Bioinformatics and Biomedicine (BIBM 2021). pp. 1424–1428 (2021)
23. Pan, S.J., Yang, Q.: A survey on transfer learning. IEEE Transactions on Knowledge and Data Engineering **22**(10), 1345–1359 (2010)
24. Pan, S.J., Tsang, I.W., Kwok, J.T., Yang, Q.: Domain adaptation via transfer component analysis. IEEE Transactions on Neural Networks **22**(2), 199–210 (2011)
25. Peng, M., Zhang, Q.: Weighed domain-invariant representation learning for cross-domain sentiment analysis. In: Proceedings of the 28th International Conference on Computational Linguistics. pp. 251–265. Barcelona, Spain (Online) (2020)
26. Peng, M., Zhang, Q., Jiang, Y.g., Huang, X.J.: Cross-domain sentiment classification with target domain specific information. In: Proceedings of the 56th Annual Meeting of the Association for Computational Linguistics. pp. 2505–2513 (2018)
27. Qu, X., Zou, Z., Cheng, Y., Yang, Y., Zhou, P.: Adversarial category alignment network for cross-domain sentiment classification. In: Proceedings of the 2019 Conference of the North American Chapter of the Association for Computational Linguistics: Human Language Technologies, Volume 1 (Long and Short Papers). pp. 2496–2508. Minneapolis, Minnesota (2019)
28. Schuster, M., Paliwal, K.K.: Bidirectional recurrent neural networks. IEEE transactions on Signal Processing **45**(11), 2673–2681 (1997)
29. Sharma, R., Bhattacharyya, P., Dandapat, S., Bhatt, H.S.: Identifying transferable information across domains for cross-domain sentiment classification. In: Proceedings of the 56th Annual Meeting of the Association for Computational Linguistics (Volume 1: Long Papers). pp. 968–978. Melbourne, Australia (2018)

30. Shen, J., Qu, Y., Zhang, W., Yu, Y.: Wasserstein distance guided representation learning for domain adaptation. In: Proceedings of the Thirty-Second AAAI Conference on Artificial Intelligence (AAAI 2018). pp. 4058–4065 (2018)
31. Shi, B., Fu, Z., Bing, L., Lam, W.: Learning domain-sensitive and sentiment-aware word embeddings. In: Proceedings of the 56th Annual Meeting of the Association for Computational Linguistics, Melbourne, Australia (ACL 2018). pp. 2494–2504 (2018)
32. Si, S., Tao, D., Geng, B.: Bregman divergence-based regularization for transfer subspace learning. IEEE Transactions on Knowledge and Data Engineering **22**(7), 929–942 (2010)
33. Sun, S., Shi, H., Wu, Y.: A survey of multi-source domain adaptation. Information Fusion **24**, 84–92 (2015)
34. Tan, C., Sun, F., Kong, T., Zhang, W., Yang, C., Liu, C.: A survey on deep transfer learning. In: Proceedings of the International Conference on Artificial Neural Networks (ICANN 2018). pp. 270–279 (2018)
35. Wang, D., Lu, C., Wu, J., Liu, H., Zhang, W., Zhuang, F., Zhang, H.: Softly associative transfer learning for cross-domain classification. IEEE Transactions on Cybernetics **50**(11), 4709–4721 (2020)
36. Wang, M., Deng, W.: Deep visual domain adaptation: A survey. Neurocomputing **312**, 135–153 (2018)
37. Xue, Q., Zhang, W., Zha, H.: Improving domain-adapted sentiment classification by deep adversarial mutual learning. In: Proceedings of the AAAI Conference on Artificial Intelligence. pp. 9362–9369 (2020)
38. Zellinger, W., Grubinger, T., Lughofer, E., Natschlager, T., Saminger-Platz, S.: Central moment discrepancy for domain-invariant representation learning. arXiv preprint arXiv: 1702.08811 (2017)
39. Zellinger, W., Moser, B.A., Grubinger, T., Lughofer, E., Natschlager, T., Saminger-Platz, S.: Robust unsupervised domain adaptation for neural networks via moment alignment. Information Science **483**, 174–191 (2019)
40. Zhang, B., Xu, X., Yang, M., Chen, X., Ye, Y.: Cross-domain sentiment classification by capsule network with semantic rules. IEEE Access **6**, 58284–58294 (2018)
41. Zhang, K., Zhang, H., Liu, Q., Zhao, H., Zhu, H., Chen, E.: Interactive attention transfer network for cross-domain sentiment classification. In: Proceedings of the 33rd AAAI Conference on Artificial Intelligence (AAAI 2019). pp. 5773–5780 (2019)
42. Zhou, J., Tian, J., Wang, R., Wu, Y., Xiao, W., He, L.: Sentix: A sentiment-aware pre-trained model for cross-domain sentiment analysis. In: Proceedings of the 28th International Conference on Computational Linguistics. pp. 568–579 (2020)
43. Zhuang, F., Cheng, X., Luo, P., Pan, S.J., , He, Q.: Supervised representation learningčžtransfer learning with deep autoencoders. In: Proceedings of the 24th International Joint Conference on Artificial Intelligence. pp. 4119–4125 (2015)
44. Ziser, Y., Reichart, R.: Pivot based language modeling for improved neural domain adaptation. In: Proceedings of the 2018 Conference of the North American Chapter of the Association for Computational Linguistics: Human Language Technologies. pp. 1241–1251 (2018)

Chapter 6
Multi-Source Domain Adaptation for Emotion Classification Using Bi-LSTM and Broad Learning

Abstract Existing domain adaptation methods for TEC task have tended to explore single-source domain, rather than on multi-source domain adaption. However, in fact, there is limited information and volume from single-source domain, which may affect emotion classification efficiency. Thus, to improve the performance of domain adaptation, we present a novel multi-source domain adaptation approach for TEC task, by combining BL and DL in this chapter. Specifically, to better capture the contextual features, we first use BERT and Bi-LSTM to extract DIF from each source domain to the same target domain. Then, to more effectively conduct the multi-label classification task, we adopt BL to train multiple classifiers on the basis of DIF. In addition, we design a co-training model to boost these classifiers. Finally, we conduct several experiments on four data sets by comparing with the baseline methods. The experimental results show that our proposed method can outperform the baseline methods for the textual emotion classification.

6.1 Introduction

Nowadays, existing TEC methods focus on exploring on the single-domain [15]. However, few of them study cross-domain emotion classification. In addition, most existing approaches for domain adaptation focus on performing CDSC task under single-source scenario. If there is a prominent difference for feature distribution between source and target domains, its performance will decline heavily, which is called "negative transfer" [1, 8, 19]. However, multi-source domain adaptation (MDA) is more feasible and valuable than that of single-source domain adaptation, which has attracted more and more attention in academia and industry.

As to MDA, Li et al. [12] investigated the domain discrepancy between pairwise sources, and provided a better bound for it. Hoffman et al. [9] investigated the MDA problem by providing the normalized solutions with strong theoretical guarantees for the cross-entropy loss and squared loss, and designing new algorithms to determine the distribution-weighted combination solution for the cross-entropy loss and squared loss.

S. Peng, L. Cao, *Textual Emotion Classification Using Deep Broad Learning*, Socio-Affective Computing 11, https://doi.org/10.1007/978-3-031-67718-2_6

Chen et al. [1] presented a GAN-based framework for MDA, which is called multi-source adaptation network (MSAN). MSAN was composed by multiple GAN architectures to learn bidirectional transformations from multi-source domains to target domain. Wright and Augenstein [16] explored the unsupervised MDA problem by using large pre-training transformer model, domain adversarial training, and mixture of experts technique. Yin et al. [20] presented a universal multi-source adaptation network (UMAN) to solve MDA task. In UMAN, a novel pseudo-margin vector and its weighted form were adopted to estimate the reliability of each known class belonging to the common label set, which helps the adversarial training better to align the distributions of multi-source domains and target domain.

As to CDSC, existing methods for CDSC may be divided as follows: single-source CDSC and multi-source CDSC. Existing methods for single-source CDSC have been discussed in Chap. 5. In this subsection, we will only introduce the existing methods for multi-source CDSC.

Wu and Huang [17] proposed a sentiment domain adaptation with multiple sources approach, which consists of two steps: training and adaptation. At the training stage, the global and domain-specific sentiment knowledge were extracted from multiple source domains. At the adaptation stage, the knowledge was transferred to target domain by combining the sentiment graph knowledge of target domain with the similarities between target and source domains.

Guo et al. [8] proposed a mixture-of-experts (MoE) approach for multi-source CDSC. They modeled the domain relations with a point-to-set distance metric. Khan et al. [10] proposed a transfer learning approach for multi-source CDSC, by revising SentiWordNet sentiment scores based on cosine similarity and using SVM-based model learning.

Zhao et al. [21] proposed a framework for the multi-source CDSC task based on MDA and joint learning. This framework uses Bi-GRU and CNN for deep feature extraction and uses soft parameter sharing for information transfer across tasks. Xu et al. [18] adopted the hierarchical adversarial neural network (HANN) for the multi-domain CDSC task. HANN not only can capture the distinct properties of each domain, but also can share common information across multiple domains.

Yang et al. [19] presented a method for multi-source CDSC by using a Granger-causal objective. Specifically, for each source domain, they built an expert model by using a sentiment-guided capsule network, which captures the domain invariant knowledge that bridges the knowledge gap between the source and the target domains. Then, they adopted an attention mechanism to assign importance weights to a MoE, each of which specializes in a different source domain. In addition, they utilized a Granger-causal objective to assign the weights to individual experts. Lin et al. [13] proposed a MDA method for visual sentiment classification, called multi-source sentiment GAN, which is based on variational autoencoders and GANs.

Dai et al. [4] designed two transfer learning frameworks for multi-source CDSC based on the adversarial learning. The first framework is called Weighting scheme based unsupervised domain adaptation (UDA) framework, which combines the source classifiers to acquire pseudo labels for target instances directly. The second framework is called two-stage training based UDA, which further exploits these pseudo labels to train a target private extractor. Zhao et al. [22] proposed an instance-level MDA framework, named curriculum cycle-consistent GAN, for CDSC with multiple-sources. It consists of three components: pre-trained text encoder, intermediate domain generator, and task classifier.

Fu and Liu [6] presented a method for the multi-source CDSC task, called contrastive transformer-based domain adaptation (CTDA). CTDA includes four stages: (1) mixed selector design: to weight all related sources or select the top-k sources according to the spatial similarity between two domains, (2) domain adaptor constructing: to extract domain-invariant features by minimizing the Wasserstein distance between two domains, (3) domain discriminator building: to capture the domain-private features via contrastive learning, and (4) classifier training: to predict the sentiment tendency of the target domain according to multiple trained source classifiers.

Despite the progress made by the above methods, there are still many challenges for MDA task. The first is that the same word may deliver different emotions in different domains. For instance, the word "long" may deliver many different emotions. In the domain of smartphone, "long" expresses a positive emotional tendency, while "long" expresses a negative emotional tendency in the domain of catering service. The second is that long-distance interdependent features are difficult to be captured effectively. The last is that most existing methods on MDA are DL-based models, which are disadvantages (e.g., GAN may fall into the problem of non-convergence and model collapse).

To address the above challenges, in this chapter, we present a novel method, called multi-source BL (MBL), for MDA-based emotion classification. Specifically, we first design a model to extract DIF from each source domain to the same target domain, by adopting BERT and Bi-LSTM. Then, on the basis of DIFs, multiple classifiers can be trained by exploiting BL with the labeled data. In addition, we design a co-training model to boost together these classifiers. Finally, extensive experiments are conducted on four data sets by comparing with the baseline methods.

6.2 Methodology

Here, we first provide problem definition for MDA-based emotion classification, and introduce Kullback Leibler (KL) [23] metric which can be applied to measure the divergence between the probability distribution of any two random variables. Then, as to MBL, we provide a description on how to obtain DIF between target domain and each source domain. At last, we further describe how to fuse them by designing a co-training algorithm. The framework of MBL is shown in Fig. 6.1.

MBL consists of three components: domain-invariant feature, domain-invariant classifier and co-training. Domain-invariant feature aims to capture the domain-invariant feature between the target and each source domain. Domain-invariant

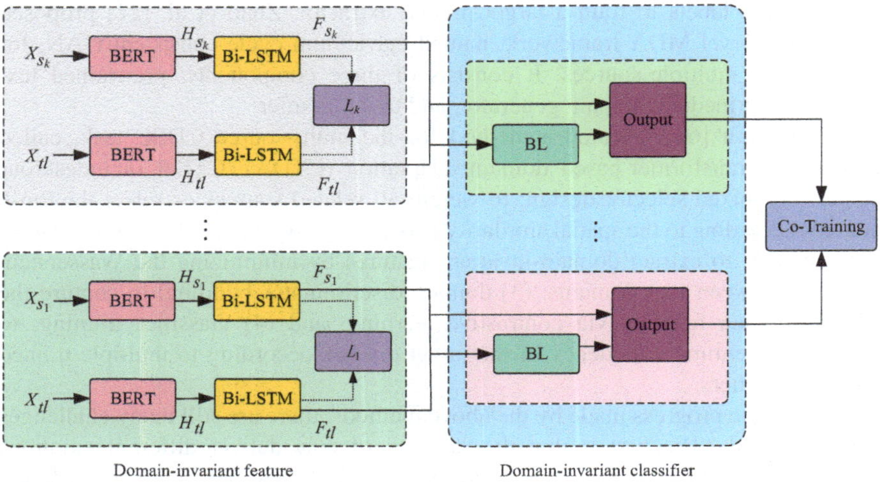

Fig. 6.1 The framework of MBL

classifier aims to classify the textual emotion. Co-training aims to train the relevant parameters.

6.2.1 Problem Definition

For MDA task, we define k source domains $D_s = \{D_{s1}, D_{s2}, \ldots, D_{sk}\}$ and a target domain D_t. Suppose that there are labeled data in the j-th source domain D_{sj}, that is $D_{sj} = \left\{x_{sj}^i, y_{sj}^i\right\}_{i=1}^{N_{sj}}$, $j = 1, 2, \ldots, k$, where N_{sj} denotes the total number of labeled data in D_{sj}. In addition, suppose that there are also a set of unlabeled data in D_t, that is $D_{tu} = \left\{x_{tu}^i\right\}_{i=1}^{N_{tu}}$, where N_{tu} denotes the total number of unlabeled data in D_t, and a few labeled data in D_t, that is $D_{tl} = \left\{x_{tl}^i, y_{tl}^i\right\}_{i=1}^{N_{tl}}$, where N_{tl} denotes the labeled data in D_t. The purpose of multi-source cross-domain emotion classification is to train k robust classifiers utilizing the labeled data in multi-source and target domains, and then to adopt them for predicting the unlabeled target data.

6.2.2 Kullback-Leibler (KL) Divergence

KL divergence is a measure of the dissimilarity between two probability distributions. In other words, KL is designed to measure the probability distribution divergence between any two random variables. There are two separate definitions of

KL divergence, one is for discrete random variables and the other is for continuous variables. In this chapter, we focus on providing the definition for discrete random variables.

Let X_s and X_t be two discrete random variables with supports two domains D_s and D_t, and probability distribution functions $P(x_s)$ and $P(x_t)$. Then the KL divergence between $P(x_s)$ and $P(x_t)$ is defined as follows:

$$\mathrm{KL}\left(P(x_s), P(x_t)\right) = \sum P(x_s) \log\left(\frac{P(x_s)}{P(x_t)}\right), \tag{6.1}$$

KL divergence $\mathrm{KL}\left(P(x_s), P(x_t)\right)$ measures how much the distribution defined by $P(x_t)$ is dissimilar from the reference distribution defined by $P(x_s)$. If two distributions are the same, KL between them is set to zero. Because a measure for the asymmetry of the two probability distributions, KL has been widely adopted to reduce the domain shift in domain adaptation. We will directly use KL divergence to compute the distance of mean value of sample feature distribution.

6.2.3 Overview of Broad Learning

The main idea of BL is that word embeddings generated by BERT are linearly mapped into m groups of feature nodes, and then feature nodes are nonlinearly mapped into n groups of enhancement nodes. Finally, feature nodes and enhancement nodes are input into the output layer for obtaining the probability distribution of emotions. During the training process of BL, the weights of feature nodes and enhancement nodes are generated randomly and fixed, and the weights of output layer are optimized by the ridge regression method.

Given a training set $\{X, Y\} \in \mathbb{R}^{N\times(D+C)}$, where N denotes the total number of samples, D denotes the dimension of word embedding, and C denotes the number of emotion types, word embeddings X are linearly mapped into m groups of feature nodes. The h-th group of feature nodes F_h is represented as follows:

$$F_h = \phi\left(XW_{fh} + \beta_{fh}\right) \in \mathbb{R}^{N\times p}, h = 1, 2, \ldots, m, \tag{6.2}$$

where ϕ denotes linear activation function, p denotes the number of feature nodes in each group, and W_{fh} and β_{fh} denote the weight and bias generated randomly, respectively.

All the linear feature matrices are concatenated by m groups of feature nodes, which is represented as follows:

$$F^m \triangleq [F_1, F_2, \ldots, F_m]. \tag{6.3}$$

Then F^m is nonlinearly mapped into n groups of enhancement nodes. The i-th group of enhancement nodes E_i is represented as follows:

$$E_i = \varphi\left(F^m W_{ei} + \beta_{ei}\right) \in \mathbb{R}^{N\times q}, i = 1, 2, \ldots, n, \tag{6.4}$$

where φ denotes nonlinear activation function (e.g., tanh, sigmoid), q denotes the total number of enhancement nodes in each group, and W_{ei} and β_{ei} denote the weight matrix and bias matrix generated randomly, respectively.

All the linear feature matrices are concatenated by n groups of enhancement nodes, which is represented as follows:

$$E^n \triangleq [E_1, E_2, \ldots, E_n]\,. \tag{6.5}$$

The actual input of BL is concatenated by F^m and E^n, which can be represented as follows:

$$A = \left[F^m, E^n\right] \in \mathbb{R}^{N\times(mp+nq)}. \tag{6.6}$$

Thus, the BL's final estimated output $\hat{Y}$ can be represented as follows:

$$\begin{aligned}\hat{Y} &= \left[g\left(XW_{f1} + \beta_{f2}\right), \ldots, g\left(XW_{fm} + \beta_{fm}\right)\right.\\ &\quad \left.|\,\varphi\left(F^m W_{e1} + \beta_{e1}\right), \ldots, \varphi\left(F^m W_{en} + \beta_{en}\right)\right]W\\ &= [F_1, F_2, \ldots, F_n\,|E_1, E_2, \ldots, E_m\,]\,W\\ &= \left[F^n, E^m\right]W\\ &= AW,\end{aligned} \tag{6.7}$$

where W denotes the weights of output layer.

Thus, according to the matrix analysis theory, W can be represented as follows:

$$W = A^+Y, \tag{6.8}$$

where A^+ denotes the pseudo inverse matrix of A.

The overall training process of BL is listed in Algorithm 6.1.

According to the analysis in Algorithm 6.1, the time complexity of BL is $O(rmp + rnq)$.

6.2.4 Feature Extraction

Since BERT is a self-supervised learning method, it conducts pre-training on large-scale unlabeled corpus by using transformer encoder structure, and converts

Algorithm 6.1 The training process of BL

Input: X
Output: W

1: **for** $k = 1; k \leq m$ **do**
2: Randomly generate W_{fk}, β_{fk}
3: Compute $F_k = \phi\left(XW_{fk} + \beta_{fk}\right)$
4: **end for**
5: Set feature nodes $F^m \triangleq [F_1, F_2, \ldots, F_m]$
6: **for** $i = 1; i \leq n$ **do**
7: Randomly generate W_{ei}, β_{ei}
8: Compute $E_i = \varphi\left(F^m W_{ei} + \beta_{ei}\right)$
9: **end for**
10: Set enhancement nodes $E^n \triangleq [E_1, E_2, \ldots, E_n]$
11: Set $A = [F^m, E^n]$
12: Compute output weight W

the distance between two words at any position into vector representation by using attention mechanism. It can effectively solve the problem of long-distance dependency in NLP, and simultaneously obtain more useful information in text. In addition, compared with LSTM, CNN, and RNN, Bi-LSTM can better capture contextual information in text. Thus, to effectively conduct MDA-based emotion classification, we adopt BERT and Bi-LSTM to extract DIF.

Firstly, we utilize BERT to generate word embeddings for source labeled data X_{sj} in j-th source domain X_{sj}, and target labeled data X_{tl}, which are represented as follows:

$$\begin{aligned} H_{sj} &= \mathrm{BERT}\left(X_{sj}; \theta_j^1\right) \in \mathbb{R}^{(N_{sj}l)\times 768}, \\ H_{tl} &= \mathrm{BERT}\left(X_{tl}; \theta_j^1\right) \in \mathbb{R}^{(N_{tl}l)\times 768}, \end{aligned} \tag{6.9}$$

where BERT denotes the representations of DIF encoded by BERT, θ_j^1 denotes the corresponding parameter, and l denotes the sequence length.

Then, on the basis of these representations, the contextual features and long-distance dependencies can be extracted by Bi-LSTM. The encoding results for j-th source domain X_{sj} and X_{tl} can be represented as follows:

$$\begin{aligned} F_{sj} &= \mathrm{BiLSTM}\left(H_{sj}; \theta_j^2\right) \in \mathbb{R}^{N_{sj}\times r}, \\ F_{tl} &= \mathrm{BiLSTM}\left(H_{tl}; \theta_j^2\right) \in \mathbb{R}^{N_{tl}\times r}, \end{aligned} \tag{6.10}$$

where BiLSTM denotes the representations of DIF encoded by Bi-LSTM, θ_j^2 denotes the corresponding parameter, and r denotes the representation dimension.

As to DIF, we hope it can encode features shared by both source and target domains. From the probability distribution view, we hope that the distributions of the mapped outputs obtained by DIF from source and target data are similar. Thus, we utilize KL regularizer onto the features of source data F_{sj} and target data F_{tl}. KL divergence can be defined as follows:

$$L_j^{sim} = \mathrm{KL}\left(F_{sj}, F_{tl}\right). \tag{6.11}$$

The distribution discrepancy between F_{sj} and F_{tl} can be reduced by minimizing the loss L_j^{sim}, which contributes to obtain DIF.

6.2.5 BL-Based Classifier

Since BL has two advantages: (1) The features can be nonlinearly mapped into high-dimensional feature space to further extract useful information; (2) It needs only to update the weights of output layer by the ridge regression, so the computational cost is very low. On the basis of textual features extracted by BERT and Bi-LSTM, BL is adopted to further capture features in texts from both source and target domains, and to design a DIC on the basis of DIF.

According to the BL theory, F_{sj} and F_{tl} are nonlinearly mapped into n groups of enhanced nodes. Thus, the i-th group of enhanced nodes for the j-th source domain can be represented as follows:

$$E_{ij} = \varphi\left(\left[F_{sj}, F_{tl}\right] W_{ei} + \beta_{ei}\right) \in \mathbb{R}^{(N_{sj}+N_{tu})\times q}, i = 1, 2, \ldots, n. \tag{6.12}$$

Let $E_j^n \triangleq \left[E_{1j}, E_{2j}, \ldots, E_{nj}\right]$ denote the matrix concatenated by n groups of enhancement nodes of the j-th source domain. Therefore, the DIC's final estimated output $\hat{Y}_j$ can be represented as follows:

$$\hat{Y}_j = \left[F_{sj}, F_{tl}, E_j\right] W_j = A_j W_j, \tag{6.13}$$

where A_j denotes all the input features of DIC, and W_j denotes the weight of output layer of DIC.

In the inference stage to predict X_{tu}, we weight and sum the output of each DIC to obtain the emotion of the unlabeled target samples, which can be represented as follows:

$$\hat{Y} = \sum_{j=1}^{k} \hat{Y}_j. \tag{6.14}$$

6.2.6 Co-Training

The co-training process of this method is divided into two steps: (i) training a encoder of DIF for each source-target domain pair; (ii) training a classifier DIC for each DIF. As to DIF, the classification loss on DIF is utilized to measure the difference between ground truth and predication of source domain and target domain. Since minimum entropy can help to constrain the predicted values of target domain samples and to increase the distance between target sample and classification decision boundary, to obtain higher confidence for the prediction results in target domain, we conduct the entropy minimization for the probability distribution of each kind of sample, and its loss function is represented as follows:

$$\begin{aligned} L_j^c = & -\frac{1}{N_{sj}+N_{tl}} \sum_{i=1}^{N_{sj}} y_{sj}^i \log Q\left(y_{sj}^i \middle| F_{sj}^i\right) \\ & - \eta_j \frac{1}{N_{sj}+N_{tl}} \sum_{m=1}^{N_{tl}} y_{tl}^m \log Q\left(y_{tl}^m \middle| F_{tl}^m\right), \end{aligned} \tag{6.15}$$

where F_{sj}^i and F_{tl}^m denote the encoding result of Bi-LSTM for the i-th source example x_{sj}^i and the m-th target example x_{tl}^m, respectively, and η_j denotes the weight of target domain loss.

Thus, the total loss function includes the difference measured by KL divergence between feature distributions between each source and target domain, and the minimum entropy loss in each source and target domain, which is represented as follow:

$$L = \sum_{j=1}^{k} (L_j^c + \delta_j L_j^{sim}), \tag{6.16}$$

where δ_j denotes the weight of loss L_j^{sim}.

As to DIC, we need to obtain an appropriate W_j to maintain the difference between Y_j and $\hat{Y}_j$ as small as possible. Thus, the ridge regression is adopted as objective function, which is represented as follows:

$$\underset{W_j}{\operatorname{argmin}} \left(\left\| Y_j - \hat{Y}_j \right\|_2^2 + \lambda \left\| W_j \right\|_2^2 \right), \tag{6.17}$$

where λ denotes the regularization parameter and Y_j denotes the actual output of DIC.

Thus, according to the regularized least square method, W_j can be represented as follows:

$$W_j = \left(\lambda I + A_j A_j^{\mathrm{T}}\right)^{-1} A_j^{\mathrm{T}} Y_j. \tag{6.18}$$

Here, we regard DIF for each source-target domain pair as independent spaces in view of target domain data. On the basis of DIF, we train DIC with respect to parameter W_j, and employ both a small number of target labeled data and source labeled data to train DIC.

Thus, through the above deducing, the algorithm of MBL is shown in Algorithm 6.2.

Algorithm 6.2 MBL algorithm

Input: T: max iteration;
 k: the total number of souce domain;
 X_{sj}: the souce domain labeled data;
 X_{tl}: the target domain labeled data;
 X_{tu}: the target domain unlabeled data;
Output: $\hat{Y}$
Training:
Set $t = 1$;
while $t \leq T$ **do**
  for each source domain labeled data $(X_{s1}, X_{s2}, \ldots, X_{sk})$ **do**
    Obtain DIF based on X_{sj} and X_{tl} by Eq. (6.9) and Eq. (6.10);
    Compute KL divergence onto between F_{sj} and F_{tl} by Eq. (6.11);
    Obtain DIC based on F_{sj} and F_{tl} by Eqs. (6.12) and (6.13);
    Train DIF by optimizing Eq. (6.16);
    Train DIC by Eq. (6.18);
    Predict X_{tu} by Eq. (6.14);
    Select p samples X_{tu}^{j} with the highest confidence from X_{tu};
  end for
  Remove samples $X_{tu}^{1} \cup X_{tu}^{2} \cup \ldots \cup X_{tu}^{k}$ from X_{tu} and add them to X_{tl};
  $t = t + 1$
end while
Testing:
Use MBL to predict X_{tu} and obtain $\hat{Y}$;

According to Algorithm 6.2, the time complexity of MBL is $O(lr + r^2 + rmp + rnq)$. In addition, the time complexity of SVM, Bi-GRU, TextCNN, DANN, and UMDA is $O(l^2)$, $O(lr+r^2)$, $O(uvlr^2)$, $O(l^2)$, and $O(l^2)$, respectively. where l denotes the sequence length, r denotes the representation dimension, v denotes the kernel size of convolution, and u denotes the number of convolution layers.

In BL, m, n, p, and q are usually set to a small value, respectively. In general, they are smaller than r, and l is far larger than m, n, p, and q. Thus, the time complexity of MBL approximately is $O(lr^2)$ and is smaller than that of SVM, TextCNN, DANN, and UMDA. In addition, the time complexity of Bi-LSTM is close to Bi-GRU.

6.3 Experiment and Analysis

6.3.1 Data Set

To verify the effectiveness of our proposed method, we conduct comparative experiments on the data sets from four domains: movie (M), hotel (H), electronics

(E), and clothing (C). Detailed information for each kind of data set has shown in Chap. 5.

According to these data, we constructed twelve cross-domain seven classification tasks. In each DA task, there are 1000 labeled source domain examples, 1000 unlabeled target domain examples, and 50 labeled target domain examples selected for training data. In addition, to effectively fine-tune the hyper-parameters, 500 target examples are selected as developing data, and the remainder examples are treated as testing data. This setting is adopted in each baseline method and MBL.

6.3.2 Baseline Method

In this subsection, many experiments are carried out by comparing our proposed MBL with the following baseline methods: SVM [5], Bi-GRU [2], TextCNN [11], DANN [7], and UMDA [20].

- **SVM**: It is a non-domain-adaptive method.
- **Bi-GRU**: It is a non-domain-adaptive method, which is implemented by adopting Bi-GRU.
- **TextCNN**: It is a convolutional neural network for emotion classification without considering contextual information of text.
- **DANN**: It is a domain-adaptive method, which utilizes the domain classifier for minimizing the discrepancy between two domains in an adversarial training manner.
- **UMDA**: It is an adversarial training method for universal multi-source adaptation.

6.3.3 Parameter Setting

In our experiments, all the word embeddings for texts are initialized to generate 768 dimension vectors. In MBL, the Chinese BERT pre-training model was adopted to generate word embeddings. Chinese BERT was presented by Cui et al. [3], which is composed of twelve transformer blocks, and is pre-trained by a large amount of Chinese corpus (including Chinese Wikipedia, news). The nodes of hidden layer for Bi-LSTM is set to 200. The enhancement nodes of BL are composed of 20 groups, 50 nodes in each group, and tanh activation function.

Model optimization was conducted by using AdamW update strategy [14] with the weight decay set to 0.01 and the initial learning rate set to 1e-5. The relevant hyperparameters to the best performance for the validation set are obtained by grid search, and the weight of the loss item is set by $\delta = 0.5$, $\eta = 1$. The factor for each iteration of co-training p is set by 5, and the regularization parameter λ is set by 0.001.

6.3.4 Main Result

To verify the effectiveness of MBL, we compare it with the baseline methods. We adopt the accuracy to evaluate these methods. The experimental results for each method are shown in Tables 6.1 and 6.2, respectively. The overall comparison for the accuracy is shown in Fig. 6.2. As to Table 6.1, C+E→H indicates that domains C and E are utilized as source domains, which is transferred to target domain H. As to Table 6.2, C+E+M→H indicates that domains C, E, and M are utilized as the source domains, which is transferred to the target domain H.

As to Tables 6.1 and 6.2, we evaluate our method over twelve transfer pairs and four transfer pairs, respectively, on a total number of 41,628 testing samples. From Tables 6.1 and 6.2, it is found that MBL can achieve consistently the best classification performance on accuracy for the data sets. Compared with SVM, Bi-GRU, TextCNN, DANN, and UMDA, MBL outperforms SVM by 28.18, 23.33, 7.76, 6.41, and 4.26% on average for two source domains, respectively, and by 25.06, 19.80, 5.93, 5.34, and 3.97% on average for three source domains, respectively.

In addition, compared with the accuracy on two source domains, the result of three source domains for each baseline method is improved by 6.01, 6.42, 4.72,

Table 6.1 The accuracy of different methods with two source domains

Source→Target	SVM	Bi-GRU	extCNN	DANN	UMDA	MBL-KL	MBL
C+E→H	0.5059	0.5975	0.7896	0.8318	0.842	0.8758	**0.9098**
C+M→H	0.6120	0.6512	0.8307	0.825	0.8871	0.9155	**0.9388**
M+E→H	0.7376	0.7927	0.8762	0.8562	0.8643	0.8764	**0.9247**
M+E→C	0.5790	0.6301	0.8172	0.7944	0.7905	0.8234	**0.8401**
M+H→C	0.5204	0.5844	0.7579	0.7803	0.7831	0.8130	**0.8236**
H+E→C	0.6128	0.6898	0.7611	0.7646	0.7768	0.7692	**0.8019**
H+E→M	0.5543	0.5862	0.7348	0.7539	0.8042	0.8013	**0.8216**
C+E→M	0.5129	0.5490	0.6867	0.7055	0.7378	0.7286	**0.7437**
C+H→M	0.5394	0.5771	0.7890	0.8103	0.8210	0.8273	**0.8570**
H+C→E	0.6197	0.6662	0.8266	0.8413	0.8625	0.8867	**0.9209**
H+M→E	0.6256	0.6556	0.8038	0.8628	0.8876	0.8960	**0.9285**
C+M→E	0.6389	0.6600	0.8354	0.8452	0.8713	0.8849	**0.9289**
Average	0.5882	0.6367	0.7924	0.8059	0.8274	0.8416	**0.8700**

Table 6.2 The accuracy of different methods with three source domains

Source→Target	SVM	Bi-GRU	TextCNN	DANN	UMDA	MBL-KL	MBL
C+E+M→H	0.7365	0.8086	0.8668	0.8811	0.8881	0.9163	**0.9392**
C+E+H→M	0.568	0.5992	0.8135	0.8297	0.8348	0.8529	**0.8751**
C+H+M→E	0.6513	0.6853	0.8577	0.8683	0.9086	0.9053	**0.9336**
H+E+M→C	0.6374	0.7104	0.8203	0.8030	0.8054	0.8247	**0.8478**
Average	0.6483	0.7009	0.8396	0.8455	0.8592	0.8748	**0.8989**

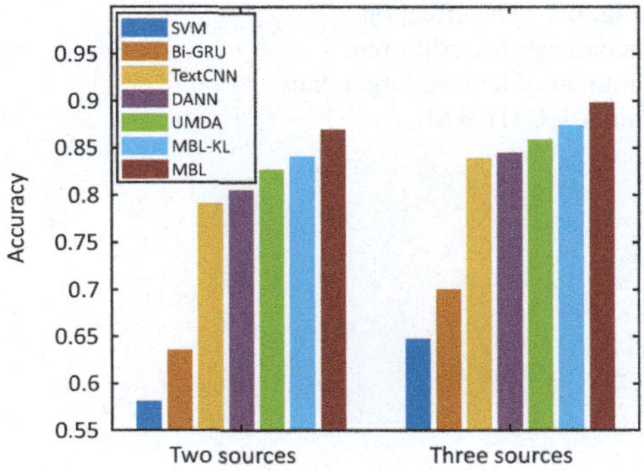

Fig. 6.2 The overall comparison for the accuracy of different methods

3.96, and 3.18%, respectively, which shows that the more source domains, the better accuracy for the cross-domain emotion classification task.

To verify the effects of domain adaptation more intuitively, we also conduct an experiment by comparing MBL with a variant without KL metric, named MBL-KL. The average accuracy of MBL-KL is 84.16 and 87.48% for two source domains and three source domains, respectively, which is lower than MBL by 2.84 and 2.41%, respectively. Through the comparison between MBL-KL and MBL, it implies that KL is important to improve the accuracy of MDA.

6.3.5 *Effect of Labeled Target Data*

To effectively analyze the effect of MBL from the number of labeled target data, we conduct the experiment via the accuracy under different amount of labeled target data on tasks C+H→M and C+H+M→E, respectively, by comparing MBL with DANN and UMDA. The specific results are shown in Figs. 6.3 and 6.4.

In Figs. 6.3 and 6.4, we provide the comparison between baseline methods UMDA, DANN and MBL under a setting that some labeled target data are randomly selected, and then mixed with the training data, and the similar trend on two kinds of transferring methods from the experimental results can be obtained in other pairs. The effect on accuracy under different amount of labeled target data on the tasks: C+H→M and C+H+M→E is shown in Figs. 6.3 and 6.4, respectively.

From Fig. 6.4, it is found that the accuracy of MBL, UMDA, and DANN is 88, 85.3, and 83%, respectively, when the number of labeled target data is zero. As the number of labeled target data increases, the accuracies of MBL, UMDA, and DANN are improved by 5, 5.8, and 4%, respectively, till the number of labeled target data is set to 50. However, the accuracies of MBL, UMDA, and DANN increase slightly, when the number of labeled target data changes from 50 to 70.

In addition, the similar phenomena can also be found in Fig. 6.3. Thus, from Figs. 6.3 and 6.4, it is shown that the difference among these three methods almost remains unchanged as the amount of labeled target data increases, and MBL can

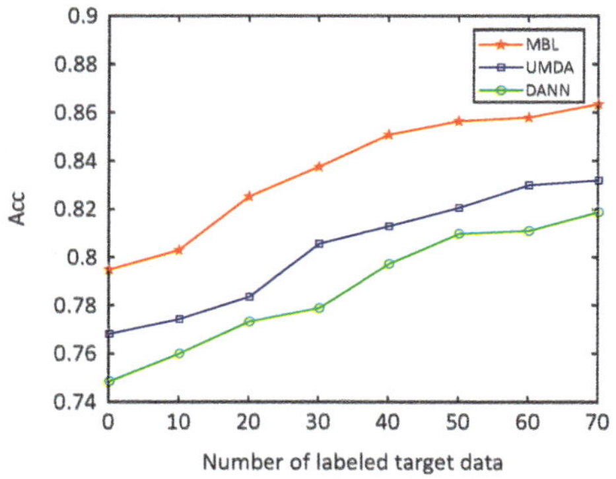

Fig. 6.3 The effect on accuracy under different amount of labeled target data on task C+H→M

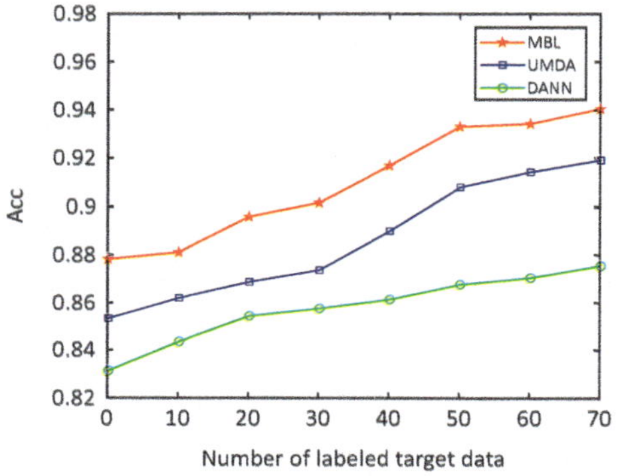

Fig. 6.4 The effect on accuracy under different amount of labeled target data on task C+H+M→E

also maintain its dominant position. It is also shown that MBL is more effective in the case of less labeled target data than the baseline methods, and can benefit from more labeled target data.

6.3.6 *Effect of Groups of Enhancement Nodes*

To effectively analyze the effect on MBL under different groups of enhancement nodes, we conduct the experiment to analyze the effect on accuracy under different amount of groups of enhancement nodes. We demonstrate that the accuracy variation for MDA-based emotion classification under different groups of enhancement nodes. The specific results are shown in Figs. 6.5, 6.6, 6.7, and 6.8.

From Figs. 6.5, 6.6, 6.7, and 6.8, it is found that the accuracy increases slightly as to most tasks, as the number of group of enhancement nodes increases. It means that the enhancement nodes can effectively improve the performance of MBL. For example, as to tasks H+M→E and C+M→E, when the group of enhancement nodes is twenty, MBL can achieve the best performance. However, the accuracy of MBL begins to decrease when the group of enhancement nodes continues to increase. The

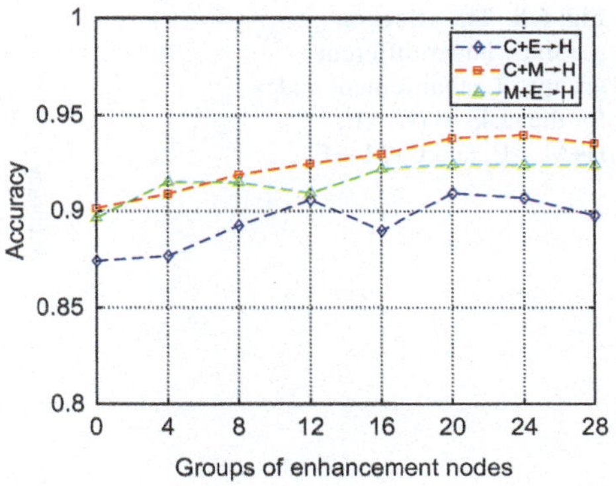

Fig. 6.5 The effect on accuracy under different groups of enhancement nodes for the tasks C+E→H, C+M→H, and M+E→H

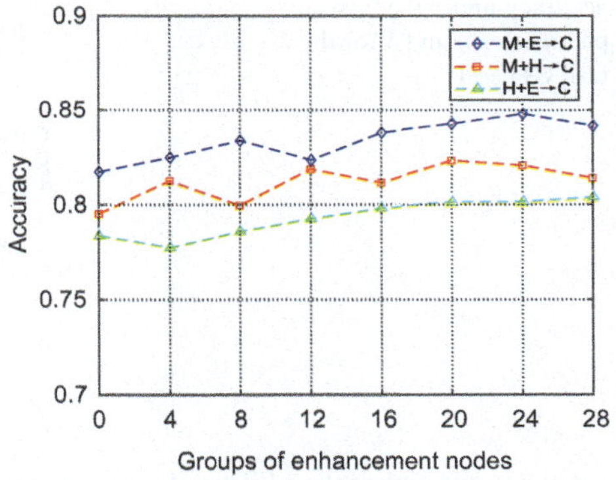

Fig. 6.6 The effect on accuracy under different groups of enhancement nodes for the tasks M+E→C, M+H→C, and H+E→C

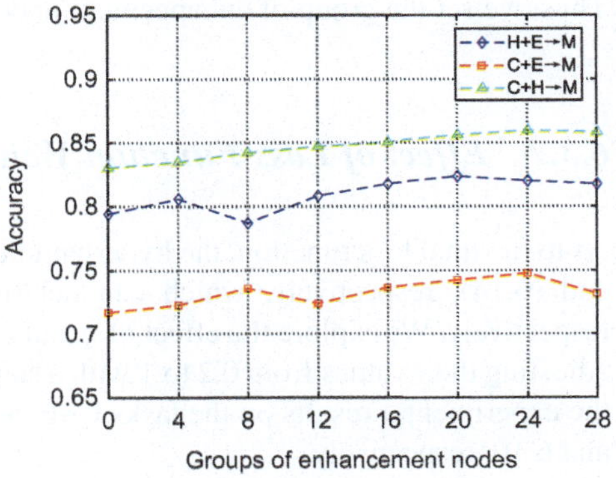

Fig. 6.7 The effect on accuracy under different groups of enhancement nodes for the tasks H+E→M, C+E→M, and C+H→M

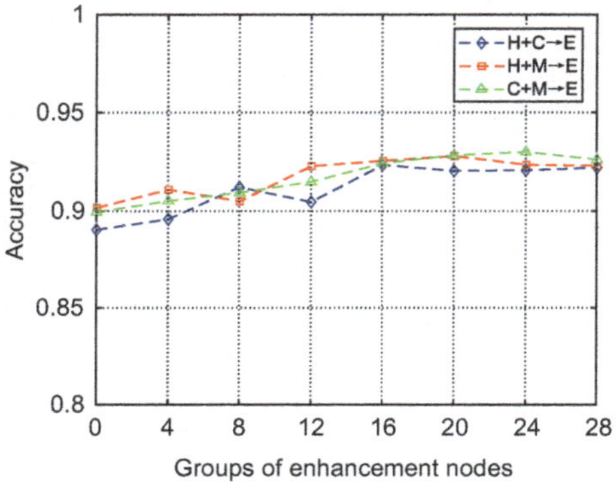

Fig. 6.8 The effect on accuracy under different groups of enhancement nodes for the tasks H+C→E, H+M→E, and C+M→E

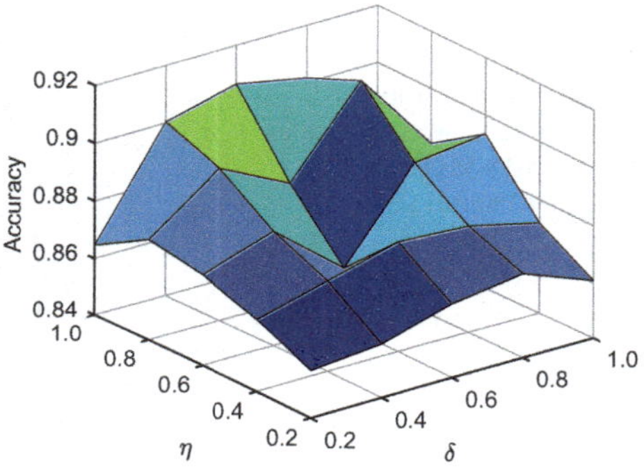

Fig. 6.9 The effect on accuracy under the parameters η and δ for the task C+E→H

main reason is that the ability of each group of enhancement nodes in BL is limited, rather than infinitely increase with the increase of group of enhancement nodes. Thus, we set the group of enhancement nodes twelve in our other experiments.

6.3.7 Effect of Loss Function Weight

As to the final loss function, the hyperparameters η and δ have defined in Eqs. (6.13) and (6.14), respectively, which can determine the importance of KL and BL, respectively. We explore the effect of η and δ on the overall performance of MBL by adjusting their values from 0.2 to 1 with a step size of 0.2, respectively. For example, the experimental results on the tasks C+E→H and C+M→E are shown in Figs. 6.9 and 6.10, respectively.

From Figs. 6.9 and 6.10, it is found that the accuracy increases as the value of η increases from 0.9 to 1. The main reason is that large target domain weights can help MBL to learn the knowledge of target domain more effectively. It is also found that MBL can achieve better classification performance when $\eta \in [0.9, 1]$ and $\delta \in [0.4, 0.6]$ for most tasks. Thus, we set $\eta = 1$ and $\delta = 0.5$ in our experiments.

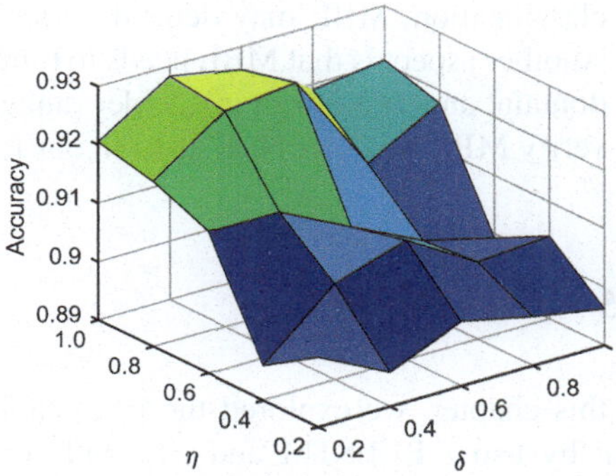

Fig. 6.10 The effect on accuracy under the parameters η and δ for the task C+M→E

Table 6.3 Accuracy comparison of BERT+Bi-LSTM+BL with BERT+CNN+BL, BERT+RNN+BL, and BERT+LSTM+BL

Source→Target	Bi-LSTM	CNN	RNN	LSTM
C+E→H	0.9098	0.9041	0.8956	0.8925
M+H→C	0.8236	0.8255	0.8101	0.8181
C+H+M→E	0.9336	0.9287	0.9264	0.9227
Average	**0.8890**	0.8861	0.8774	0.8778

6.3.8 Superiority of BERT+Bi-LSTM+BL

To verify the superiority of BERT+Bi-LSTM+BL for extracting features, we conduct an experiment by comparing MBL (i.e., BERT+Bi-LSTM+BL) with BERT+CNN+BL, BERT+RNN+BL, and BERT+LSTM+BL, for the tasks C+E→H, M+H→C, and C+M+H→E. The specific results are shown in Table 6.3.

From Table 6.3, it is found that the accuracy of BERT+Bi-LSTM+BL outperforms BERT+CNN+BL, BERT+RNN+BL, and BERT+LSTM+BL in most tasks. Thus, we adopt BERT and Bi-LSTM to extract textual features in MBL.

6.4 Discussion

We provide a brief analysis on the limitations of our proposed MBL, which includes the following two aspects.

1. To validate the model proposed in this chapter, we have constructed four different domain data sets in Chinese for MDA-based emotion classification, and conducted extensive experiments on it. In comparison with the baseline methods and relevant parameters, MBL can achieve good results on the constructed cross-domain data sets. Because of a lack of English data set for MDA-based emotion classification, it fails to verify the performance and adaptation of MBL on English data set.
2. MBL may be regarded as a combining approach with feature-based and model-based. Though it is able to utilize Chinese data set for MDA-based emotion

classification, MBL may depend on specific data and network models heavily. Another aspect is that MBL needs to train model with numerous data in the target domain, and to perform in a wider range of NLP tasks. Thus, we need to further verify MBL by other NLP tasks in our future work.

6.5 Summary

In this chapter, we explored the practicability of MDA-based emotion classification by using Bi-LSTM and BL, and conducted extensive experiments on four different domain data sets collected from the public E-commerce platforms. The experimental results showed that MBL is able to reach a performance of 87.00 and 89.89% via two and three source domains, respectively, in terms of accuracy, which is a significant improvement over the baselines. In comparison with most existing methods, our proposed MBL can effectively improve the performance of TEC by using BERT, KL metric, and Bi-LSTM to extract DIFs, and the combination of Bi-LSTM and BL could be beneficially employed in MDA task. In addition, we found that the performance of MBL is better than that of SBL. In our future work, we plan to combine other DL models and BL for MDA task, and to solve the multi-modal cross-domain emotion classification task.

References

1. Chen, C., Xie, W., Wen, Y., Huang, Y., Ding, X.: Multiple-source domain adaptation with generative adversarial nets. Knowledge-Based Systems **199**, 105962 (2020)
2. Cho, K., van Merrienboer, B., Gülçehre, Ç., Bahdanau, D., Bougares, F., Schwenk, H., Bengio, Y.: Learning phrase representations using RNN encoder-decoder for statistical machine translation. In: Proceedings of the 2014 Conference on Empirical Methods in Natural Language Processing (EMNLP 2014). pp. 1724–1734. Doha, Qatar (2014)
3. Cui, Y., Che, W., Liu, T., Qin, B., Yang, Z.: Pre-training with whole word masking for chinese bert. IEEE/ACM Transactions on Audio, Speech, and Language Processing pp. 1–8 (2021)
4. Dai, Y., Liu, J., Ren, X., Xu, Z.: Adversarial training based multi-source unsupervised domain adaptation for sentiment analysis. In: Proceedings of Thirty-Fourth AAAI Conference on Artificial Intelligence (AAAI 2020). pp. 7618–7625 (2020)
5. Fan, R., Chang, K., Hsieh, C., Wang, X., Lin, C.: LIBLINEAR: A library for large linear classification. Journal of machine learning research **9**, 1871–1874 (2008)
6. Fu, Y., Liu, Y.: Contrastive transformer based domain adaptation for multi-source cross-domain sentiment classification. Knowledge-Based Systems **245**, 108649 (2022)
7. Ganin, Y., Ustinova, E., Ajakan, H., Germain, P., Larochelle, H., Laviolette, F., Marchand, M., Lempitsky, V.: Domain-adversarial training of neural networks. The journal of machine learning research **17**(1), 2030–2096 (2016)
8. Guo, J., Shah, D.J., Barzilay, R.: Multi-source domain adaptation with mixture of experts. In: Proceedings of the 2018 Conference on Empirical Methods in Natural Language Processing. pp. 4694–4703. Brussels, Belgium (2018)

9. Hoffman, J., Mohri, M., Zhang, N.: Algorithms and theory for multiple-source adaptation. In: Proceedings of the 32nd International Conference on Neural Information Processing Systems. pp. 8256–8266 (2018)
10. Khan, F.H., Qamar, U., Bashir, S.: Enhanced cross-domain sentiment classification utilizing a multi-source transfer learning approach. Soft Computing **23**(14), 5431–5442 (2019)
11. Kim, Y.: Convolutional neural networks for sentence classification. In: Proceedings of the 2014 Conference on Empirical Methods in Natural Language Processing (EMNLP). pp. 1746–1751. Doha, Qatar (2014)
12. Li, Y., Murias, M., Major, S., Dawson, G., Carlson, D.E.: Extracting relationships by multi-domain matching. In: Proceedings of the 32nd International Conference on Neural Information Processing Systems. pp. 6799–6810 (2018)
13. Lin, C., Zhao, S., Meng, L., Chua, T.: Multi-source domain adaptation for visual sentiment classification. In: Proceeding of the Thirty-Fourth AAAI Conference on Artificial Intelligence (AAAI 2020). pp. 2661–2668 (2020)
14. Loshchilov, I., Hutter, F.: Fixing weight decay regularization in adam. In: Proceedings of International Conference on Learning Representations (2018)
15. Peng, S., Cao, L., Zhou, Y., Ouyang, Z., Yang, A., Li, X., Jia, W., Yu, S.: A survey on deep learning for textual emotion analysis in social networks. Digital Communications and Networksorks **8**(5), 745–762 (2022)
16. Wright, D., Augenstein, I.: Transformer based multi-source domain adaptation. In: Proceedings of the Conference on Empirical Methods in Natural Language Processing. pp. 7963–7974 (2020)
17. Wu, F., Huang, Y.: Sentiment domain adaptation with multiple sources. In: Proceedings of the 54th Annual Meeting of the Association for Computational Linguistics. pp. 301–310. Berlin, Germany (2016)
18. Xu, Z., von Ritter, L., Serra, G.: Hierarchical adversarial training for multi-domain adaptive sentiment analysis. In: Complex Pattern Mining. vol. 880, pp. 17–32 (2020)
19. Yang, M., Shen, Y., Chen, X., Li, C.: Multi-source domain adaptation for sentiment classification with granger causal inference. In: Proceedings of the 43rd International ACM SIGIR conference on research and development in Information Retrieval. pp. 1913–1916 (2020)
20. Yin, Y., Yang, Z., Hu, H., Wu, X.: Universal multi-source domain adaptation for image classification. Pattern Recognition **121**, 108238 (2022)
21. Zhao, C., Wang, S., Li, D.: Multi-source domain adaptation with joint learning for cross-domain sentiment classification. Knowledge-Based Systems **191**, 105254 (2020)
22. Zhao, S., Xiao, Y., Guo, J., Yue, X., Yang, J., Krishna, R., Xu, P., Keutzer, K.: Curriculum cyclegan for textual sentiment domain adaptation with multiple sources. In: Proceedings of the International World Wide Web Conference (WWW 2021). pp. 541–552. Ljubljana, Slovenia (2021)
23. Zhu, Q., Bi, W., Liu, X., Ma, X., Li, X., Wu, D.: A batch normalized inference network keeps the KL vanishing away. In: Proceedings of the 58th Annual Meeting of the Association for Computational linguistics. pp. 2636–2649 (2020)

Chapter 7
Emotion Classification in Textual Conversations Using Deep Broad Learning

Abstract In this chapter, we will explore emotion classification in textual conversations (ECTC). ECTC is one of the most important tasks for NLP in recent years. However, it is a challenging task for machines to conduct emotion classification in textual conversations, because emotions rely heavily on textual context. To address the challenge, we propose a method to classify emotion in textual conversations based on DL and BL. It aims to provide a better solution to capture the local contextual information (i.e., utterance-level) in an utterance, as well as the global contextual information (i.e., speaker-level) in a conversation by using CNN, Bi-LSTM, and BL. It is the first effort to combine DL and BL to classify emotion in textual conversations.

7.1 Introduction

ECTC aims to classify the emotion of each utterance from textual conversations. It is becoming one of the most important tasks in NLP because of its wide range of applications [20], such as public opinion monitoring, mental health, behavior analysis, recommendation system, etc. In addition, it can also be applied like human-machine dialogue system, question-answering system, chat robot, and so on. Existing works for ECTC task usually capture the contextual feature by using DL models, which can be divided into sequence-based method and graph-based method [17].

As to the sequence-based method, many researchers continuously explore some new methods for ECTC task. Ma et al. proposed a multi-view network (MVN) [14] by utilizing the attention mechanism and Bi-GRU to capture the word- and utterance-level dependencies for real-time emotion recognition in conversations. As to the word-level, the context and query are regarded as the word sequences and then the word-level dependencies among utterances are modeled by using the self-attention and cross-attention. As to the utterance-level, each utterance representation is extracted separately and then the utterance-level dependencies in the context is modeled by using Bi-GRU.

S. Peng, L. Cao, *Textual Emotion Classification Using Deep Broad Learning*, Socio-Affective Computing 11, https://doi.org/10.1007/978-3-031-67718-2_7

Jiao et al. proposed a hierarchical gated recurrent unit (HiGRU) [7] framework. It exploits a lower-level GRU to model word-level inputs and employed an upper-level GRU to capture contexts of utterance-level embeddings based on the attention mechanism and Bi-GRU. Gao et al. proposed an emotion shift detection model for emotion recognition in conversations (ESD-ERC) [4]. It exploits a shared BERT-based encoder to extract the unified emotion semantic representations, utilized the Bi-LSTM to obtain the emotion shift representations, and fed the emotion semantic representations concatenated with the emotion shift representations into the context level Transformer with positional encoding for ECTC.

Ghosal et al. proposed a framework, named COSMIC [5], which utilizes GRU and RoBERTa to model various aspects of commonsense knowledge by considering mental states, events, actions, and cause-effect relations for emotion recognition in conversations. Hu et al. proposed a contextual reasoning network (DialogueCRN) [6] for ECTC task. It adopts LSTM to capture the situation-level and speaker-level context and employed the attention mechanism and LSTM to design multi-turn reasoning modules for extracting and integrating the emotional clues.

Li et al. designed HiTransformer [11] by utilizing BERT and Transformer to design two-level transformers: the lower-level transformer aims to model the word-level input and obtains the individual utterance embedding; the upper-level transformer aims to capture the contextual information and obtains utterance-level embeddings. Li et al. proposed HiTrans [10] with two transformers, which utilized the low-level transformer to generate the local utterance representations, and employed the high-level transformer to capture the global context information in conversations.

Lu et al. [13] proposed an iterative emotion interaction network to model the emotion interaction based on the attention mechanism and Bi-GRU. Zhu et al. proposed a topic-driven and knowledge-aware transformer (TODKAT) [21] to design a topic-augmented language model (LM) with an additional layer specialized for topic detection. Then, it combined the LM with the commonsense statements derived from a knowledge base on the basis of the dialogue contextual information. Finally, it introduced a transformer-based encoder-decoder to fuse the topical and commonsense information, and to process the emotion label sequence prediction.

Shen et al. designed DialogXL [16] by introducing XLNet [18] with two improvements to encode a long historical context with hundreds of words and to model the intra- and inter-speaker dependencies of different parties. Kang et al. [8] designed a semi-supervised method to help the existing ECTC models to leverage unlabeled data based on the context-augmented auxiliary training task. Feng et al. designed a speaker influence aware neural network model (SINN). It [3] adopts two GRUs to model the self-influence of each speaker separately and utilized a hierarchical matching network to comprehensively obtain the inter-speaker influence.

In the above mentioned methods, Bi-LSTM, RNN, attention, and GRU were usually utilized to capture contextual information among utterances. However, these methods neglect the relationship between utterance and the speaker. As illustrated in Fig. 7.1, there may be multiple utterances and speakers in a conversation, and

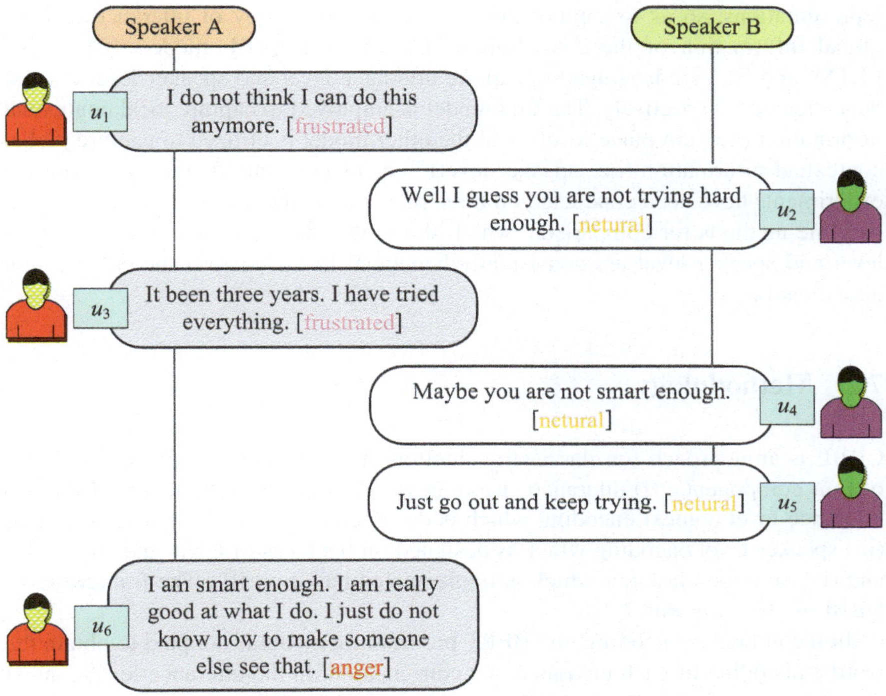

Fig. 7.1 An simple dialogue from the IEMOCAP data set

the ECTC model needs to classify the emotion of each utterance from these speakers. Thus, different from the non-conversational TEC task, the emotion of an utterance usually depends on the context of the whole conversation (i.e., long-distance contextual dependency).

Despite progress made by the above-mentioned methods, there are still many challenges for ECTC task [7]. The first is that the same word may deliver different emotions in different contexts, for instance, the word "Yes" may deliver many different emotions, such as joy, sad, and neutral. Thus, to identify the emotion of a speaker precisely, it needs to capture contextual information in textual conversation more effectively. The second is that it is difficult to capture the long-distance contextual dependency. The last is that most existing works for ECTC are based on deep learning models due to some disadvantages, e.g., RNN is difficult to model long-distance dependency.

To address these challenges, we propose a novel method for ECTC task, by integrating the advantages of CNN, Bi-LSTM, and BL, namely CBBL. It aims to provide a more effective solution to capture local contextual information (i.e., utterance-level) in an utterance, as well as global contextual information (i.e., speaker-level) in each conversation, based on CNN, Bi-LSTM, and BL. In CBBL, a popular pre-training language model BERT is utilized to obtain the initial word representations, so as to capture the contextual dependency of utterance and the global information of the conversation. Then two deep BL models (CNN, Bi-LSTM, and BL) are leveraged to capture utterance-level and speaker-level context dependencies, respectively. The first model is employed to capture local contextual information (i.e., utterance-level), and the other model is utilized to capture global contextual information (i.e., speaker-level) in a conversation. At last, the extensive experiments have been conducted on three public textual conversation data set with baseline methods for comparison, which shows that the context in both utterance-level and speaker-level are consistently beneficial to the performance of emotion classification.

7.2 Methodology

CBBL is an approach for classifying emotion in textual conversations. It consists of four components: (i) utterance encoding which is designed by using BERT; (ii) utterance-level context encoding which is designed based on CNN and Bi-LSTM; (iii) speaker-level encoding which is designed on the basis of CNN and Bi-LSTM; and (iv) emotion classifier which is implemented based on BL. The framework of CBBL is shown in Fig. 7.2.

In the utterance encoding, the BERT pre-training model is adopted to obtain the word embedding of each utterance in a conversation. In the utterance-level context encoding and speaker-level encoding, CNN and Bi-LSTM are exploited to obtain two different types of sequence information. Finally, these features are input into the emotion classifier by using BL to obtain the prediction results.

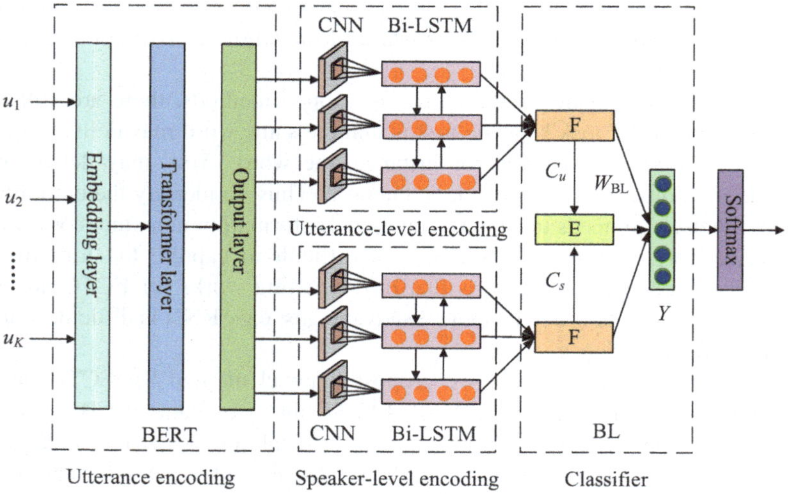

Fig. 7.2 The framework of the proposed method CBBL, E denotes enhancement nodes, F denotes feature nodes, $u_1, u_2, \ldots, u_N$ denote utterances in a conversation

7.2.1 Problem Definition

To effectively model ECTC task, we may define a textual conversation as $U = \{u_1, u_2, \ldots, u_N\}$, where U denotes the utterance set in a conversation, and N denotes the amount of utterances in a conversation; and we also define the speaker set in a conversation as $S = \{s_1, s_2, \ldots, s_M\}$, where M denotes the amount of speakers. Each utterance u_i is uttered by the speaker $s_{\phi(u_i)} \in S$, where σ is employed to map the utterance index into the corresponding speaker.

7.2.2 Utterance Encoding

We employ the Chinese BERT [2] to extract the features of utterances and to obtain word vectors, because BERT not only captures the contextual information, but also extracts the key features in text. Given N utterances in a conversation, we insert a token CLS before each utterance u_i, and then a sequence $\{CLS_1, u_1, CLS_2, u_2, \ldots, CLS_K, u_N\}$ is obtained.

Thus, we utilize the last hidden layer of BERT to obtain the vector representation of each word in each utterance u_i. If there are n_i words in an utterance u_i, we can obtain the embedding of a conversation $X = \left\{x_{CLS_1}, x_{u_1}, \ldots, x_{CLS_N}, x_{u_N}\right\}$, where $x_{u_i} \in \mathbb{R}^{n_i \times 768}$ denotes the representations of all the word embeddings in u_i.

There is a length limitation (512 tokens) in BERT input. If the total length for all the utterances in a conversation exceeds 512, it cannot deal with all the utterances in a conversation simultaneously. To address this problem, by following the solution proposed in HiTrans [10], the utterances in an overlong conversation are split into chunks whose lengths are less than 512. For example, assuming that there are 4 utterances $\{u_1, u_2, u_3, u_4\}$ in a conversation, the length of u_1 and u_2 is no more than 512, but it will exceed 512 when u_3 is added. Thus, u_1 and u_2 will be taken as a chunk and will be input into BERT, and then check whether the length of u_3 and u_4 together exceeds 512 or not. If not, u_3 will be input into BERT; otherwise, u_3 and u_4 together will be input into BERT.

7.2.3 Utterance-Level Context Encoding

In utterance-level context encoding, it aims to consider the emotional change of utterances through the utterance-dependency, which means that it can be affected by the current context and other utterances from other speakers. Since each utterance is input into BERT to extract its features separately, it is unrelated to each other. Thus, CNN and Bi-LSTM are utilized to capture the k-gram feature and long-distance dependency of the utterances, respectively. First, CNN is employed to extract the features of k-gram features and salient features by convolution and max-pooling operation; Then, Bi-LSTM is adopted to learn the correlation among each utterance.

A conversation X embeddings are input into CNN. The output of the convolutional layer can be represented as follows:

$$G_u = \sigma\left(W_u \otimes X + b_u\right) \in \mathbb{R}^{N \times q}, \tag{7.1}$$

where σ denotes the activation function, W_u and b_u denote the weights and bias of convolutional kernel, respectively, q denotes the output dimension of CNN, and $\otimes$ denotes the convolutional operation.

After the input data is extracted by the convolutional layer, the max-pooling operation is adopted to extract the salient features and reduce the dimension of the convolutional output. Thus, The output of max-pooling layer can be represented as follows:

$$P_u = \text{MaxPool}\left(G_u\right) \in \mathbb{R}^{N \times r}, \tag{7.2}$$

where MaxPool denotes the max-pooling operation of CNN and r denotes the output dimension of the max-pooling layer.

Because CNN only considers the current input, and ignores the previous degenerate information, it is difficult to capture the related contextual information among utterances. Thus, Bi-LSTM is utilized to obtain the contextual features. CNN features of each utterance $P_u = \left\{p_i^u\right\}_{i=1}^N$ are fed into Bi-LSTM to learn the correlative information among utterances:

$$\overrightarrow{c_i^u}, \overrightarrow{h_i^u} = \overrightarrow{\text{LSTM}_u}\left(p_i^u, \overrightarrow{h_{i-1}^u}\right), i = 1, 2, \ldots, N, \tag{7.3}$$

$$\overleftarrow{c_i^u}, \overleftarrow{h_i^u} = \overleftarrow{\text{LSTM}_u}\left(p_i^u, \overleftarrow{h_{i-1}^u}\right), i = 1, 2, \ldots, N, \tag{7.4}$$

where $\overrightarrow{c_i^u}$ and $\overleftarrow{c_i^u}$ denote the contextual features of two direction of i-th utterance, respectively, $\overrightarrow{h_i^u}$ and $\overleftarrow{h_i^u}$ denote the working memory of the forward LSTM and backward LSTM, respectively, and $\overrightarrow{\text{LSTM}_u}$ and $\overleftarrow{\text{LSTM}_u}$ denote the forward calculation and backward calculation of LSTM, respectively.

These two features $\overrightarrow{c_i^u}$ and $\overleftarrow{c_i^u}$ are concatenated into $c_i^u = \left[\overrightarrow{c_i^u}; \overleftarrow{c_i^u}\right]$ to generate the utterance-level feature for the i-th utterance. Thus, the utterance-level feature of a conversation can be represented as follows:

$$C_u = \left[c_1^u, c_2^u, \ldots, c_K^u\right] \in \mathbb{R}^{N \times (2l)}, \tag{7.5}$$

where l denotes the dimension of hidden state of LSTM.

7.2.4 Speaker-Level Context Encoding

In speaker-level context encoding, it aims to consider not only the emotional change between two utterances, but also self-dependency utterance between two adjacent utterances of the same speaker.

Let X be the CNN input. The output of convolutional layer can be represented as follows:

$$G_s = \varphi \left(W_s \otimes X + b_s \right) \in \mathbb{R}^{N \times q}, \tag{7.6}$$

where φ denotes the activation function, W_s denotes the weight of the convolutional layer, and b_s denotes the bias of the current convolutional layer.

Then, the output of convolutional layer is input into the pooling layer. The specific description can be represented as follows:

$$P_s = \text{MaxPool} \left(S_s \right) \in \mathbb{R}^{N \times r}. \tag{7.7}$$

The CNN features of each utterance $P_s = \left\{ p_j^s \right\}_{j=1}^{N}$ are fed into Bi-LSTM to learn the correlation of each utterance from the same speaker. The specific description can be represented as follows:

$$\overrightarrow{c_j^s}, \overrightarrow{h_{\theta,j}^s} = \overrightarrow{\text{LSTM}_s} \left(p_j^s, \overrightarrow{h_{\theta,j-1}^s} \right), j = [1, |U_\theta|], \tag{7.8}$$

$$\overleftarrow{c_j^s}, \overleftarrow{h_{\theta,j}^s} = \overleftarrow{\text{LSTM}_s} \left(p_j^s, \overleftarrow{h_{\theta,j-1}^s} \right), j = [1, |U_\theta|], \tag{7.9}$$

where $\theta = \phi \left(u_i \right)$, U_θ refers to all utterances of the speaker s_θ, $\overrightarrow{h_{\theta,j}^s}$ and $\overleftarrow{h_{\theta,j}^s}$ denote the working memory of the forward LSTM and backward LSTM, and $\overrightarrow{c_j^s}$ and $\overleftarrow{c_j^s}$ denote the contextual features of two directions of the i-th utterance.

The two features $\overrightarrow{c_j^s}$ and $\overleftarrow{c_j^s}$ are concatenated into $c_j^s = \left[\overrightarrow{c_j^s}; \overleftarrow{c_j^s} \right]$ to generate the speaker-level feature for the i-th utterance. Thus, the speaker-level feature in a conversation can be represented as follows:

$$C_s = \left[c_1^s, c_2^s, \ldots, c_M^s \right] \in \mathbb{R}^{N \times (2l)}. \tag{7.10}$$

7.2.5 Emotion Classifier

After the utterance-level and speaker-level encoding are obtained in Sects. 7.4 and 7.5, respectively, BL is adopted to calculate the weight of each emotion label and to obtain the emotion label prediction in each utterance. Since the general BL is composed of feature nodes, enhancement nodes, and an output layer, it should first linearly transform the input tensors into its feature nodes. However, the deep features of utterances were extracted through CNN and Bi-LSTM. Thus, as to BL, we directly treat these features as the feature nodes, which are nonlinearly transformed into enhancement nodes. Finally, feature nodes and enhancement nodes are concatenated to input into the output layer for calculating the weight of each label.

The utterance-level features C_u and speaker-level features C_s in a conversation are concatenated and nonlinearly mapped into m groups of enhancement nodes. The k-th group of enhancement nodes H_k can be represented as follows:

$$H_k = \xi([C_u, C_s]W_{hk} + \beta_{hk}) \in \mathbb{R}^{N\times t}, k = 1, \ldots, m, \tag{7.11}$$

where t denotes the number of enhancement nodes of each group, W_{hk} and β_{hk} are randomly generated, which denote the weight matrix and bias matrix, respectively, and ξ denotes a nonlinear activate function.

Then m groups of enhancement nods can be represented as follows:

$$H = [H_1, H_2, \ldots, H_m]. \tag{7.12}$$

Thus, the final estimated output $\hat{Y}$ can be represented as follows:

$$\begin{aligned} \hat{Y} &= [C_u, C_s, H]\, W_{\mathrm{BL}} \\ &= A W_{\mathrm{BL}}, \end{aligned} \tag{7.13}$$

where W_{BL} denotes the output weight of BL, and A denotes the actual input of BL.

To shorten the calculation time and to prevent over-fitting, the ridge regression is adopted as objective function in original BL, which is represented as follows:

$$\underset{W_{\mathrm{BL}}}{\arg\min} \left\| Y - \hat{Y} \right\|_2^2 + \lambda \left\| W_{\mathrm{BL}} \right\|_2^2 . \tag{7.14}$$

Finally, according to the regularized least square method, W_{BL} can be represented as follows:

$$W_{\mathrm{BL}} = (\lambda I + AA^{\mathrm{T}})^{-1} A^{\mathrm{T}} Y, \tag{7.15}$$

where λ denotes the regularization parameters, I denotes the identity matrix, and Y denotes the actual output of BL.

The specific process is shown in Algorithm 7.1.

Algorithm 7.1 The learning process of CBBL

Input: $U = \{u_1, u_2, \ldots, u_N\}$
Output: W_{BL}
1: Generate utterance embedding by BERT;
2: **while** stopping criterion is not satisfy **do**
3: Generate utterance-level features by Eqs. (7.3) and (7.4);
4: Generate speaker-level features by Eqs. (7.8) and (7.9);
5: Generate enhancement nodes by Eq. (7.11);
6: **end while**
7: Compute W_{BL} by Eq. (7.15);

7.3 Experiment and Analysis

In this section, we will provide a brief description for the data set, baseline methods, parameter settings, and evaluation metrics utilized in our experiments, and compare our proposed method with baseline methods. The experimental device is a laptop with Intel-i5-9400 2.9 GHz CPU and 16 GB memory, and server with Nvidia V100 GPU.

7.3.1 Data Set and Evaluation Metric

To verify the effectiveness of CBBL, the following three benchmark data sets are applied in our experiments, i.e., MELD [15], EmoryNLP [19], and IEMOCAP [1].

MELD: It is collected from *Friends* TV series. Its emotion labels include neutral, surprise, fear, sadness, joy, disgust, and anger.

EmoryNLP: It is also collected from *Friends* TV series, including neutral, joyful, scared, mad, sad, powerful, and peaceful.

IEMOCAP: It contains video, audio and text transcriptions, which is annotated with six kinds of emotion labels: neutral, happiness, sadness, anger, frustrated, and excited.

In terms of the evaluation metric in our experiments, we evaluate the overall performance by using weighted-average $F1$ (W-$F1$) score. The specific statistics of these data sets are listed in Table 7.1.

Table 7.1 Descriptions for the number of the training, validation and test data in three kinds of data sets

Data set	Conversation (Train/Val/Test)	Utterance (Train/Val/Test)	Classes
MELD	1038/114/280	9989/1109/2610	7
EmoryNLP	659/89/79	7551/954/984	7
IEMOCAP	100/20/31	4810/1000/1523	6

7.3.2 Implementation Detail

CBBL is implemented with BERT, which is composed of twelve Transformer blocks and is pre-trained on a large number of English corpus (e.g., Wikipedia, news). The region size of filters for CNN is set to two, three, and four, and its feature maps of each filter is set to 100. The nodes of hidden layer for Bi-LSTM is set to 200. The BL classifier was implemented with ten groups of enhancement nodes, fifty nodes in each group, and tanh activation functions. Model optimization was conducted using AdamW update strategy [12] with initial learning rate set to 8e-6 and weights decay set to 0.01. The hyperparameters corresponding to the best performance on the validation set are obtained by the grid search, and the regularization parameter is set to $\lambda=0.001$.

7.3.3 Baseline Method

We compare the proposed method CBBL with the following baseline methods in our experiments:

- **TextCNN** [9]: It is a convolutional neural network for the emotion classification without considering the utterance-level and speaker-level contextual information of textual conversation.
- **HiTrans**: It is a transformer-based context and speaker-sensitive model for the ECTC task.
- **HiGRU**: It is a lower-level GRU and an upper-level GRU for the ECTC task by modeling the word-level inputs and capturing the contexts of utterance-level embeddings.
- **DialogXL**: It is a XLNet-based method for conversational emotion classification by capturing useful intra- and inter-speaker dependencies.
- **MVN**: It is a multi-view network for word- and utterance-level dependencies capturing.
- **DialogueCRN**: It is the contextual reasoning networks for the ECTC task by using the attention mechanism and LSTM.
- **Lu's**: It is an iterative emotion interaction network for the ECTC task by using the attention mechanism and Bi-GRU.
- **CBBL**: It is our proposed method, which is a deep broad learning method for emotion classification by considering utterance-level and speaker-level contextual information of textual conversation.

7.3.4 *Performance Comparison*

We evaluate the overall performance by using the weighted-average $F1$ (W-$F1$), and adopt the results of the baseline methods like TextCNN and HiTrans reported in Ref. [10], the results of HiGRU and DialogXL reported in Ref. [16], the results of MVN, DialogueCRN and Lu's reported in Refs. [6, 14], and [13], respectively.

To verify the effectiveness of CBBL, we compare it with the baseline methods on MELD, EmoryNLP, and IEMOCAP, and the experimental results are listed in Table 7.2.

In Table 7.2, the experimental results are provided on MELD. It is observed that MVN can achieve the best overall performance of 62.47% on W-$F1$ score among the baseline methods. By comparison, the performance of our proposed CBBL outperforms MVN by 1.56%. CBBL can improve the overall performance by comparing with the baseline methods.

In Table 7.2, the experimental results are also provided on EmoryNLP. Among all the baseline methods, HiTrans achieves the best overall performance of 36.75% on W-$F1$. By comparison, the performance of our proposed CBBL outperforms HiTrans by 0.24%, which can prove the effectiveness of CBBL. It is shown that there is a better capability of CBBL to classify emotion in textual conversations.

From Table 7.2, it is found that MVN has the best overall performance among all the baselines. In particular, DialogueCRN, MVN and DialogXL perform better than our proposed method on IEMOCAP, which may be attributed to the capturing useful word- and utterance-level dependencies, and intra- and inter-speaker dependencies, respectively.

7.3.5 *Model Analysis*

To verify the effectiveness of our proposed method, we carry out two extensive experiments: one is applied to analyze the impact of CNN and Bi-LSTM on CBBL; the other is applied to analyze the impact of BL on CBBL. The experimental results are shown in Table 7.3.

Table 7.2 Analysis of the effectiveness with W-$F1$ for different methods on MELD, EmoryNLP, and IEMOCAP

Methods	MELD	EmoryNLP	IEMOCAP
TextCNN	52.55	27.85	43.25
HiTrans	61.94	36.75	64.5
HiGRU	56.92	31.88	59.79
DialogXL	62.41	34.73	65.94
MVN	62.47	–	65.99
DialogueCRN	58.39	–	**66.38**
Lu's	60.72	–	64.37
CBBL	**64.03**	**36.99**	62.53

Table 7.3 Analysis of the effectiveness with W-$F1$ for CBBL on MELD, EmoryNLP, and IEMOCAP

Model	MELD	EmoryNLP	IEMOCAP
BERT	60.44	34.92	54.15
BERT+CNN+Bi-LSTM	62.76	35.48	61.59
CBBL	**64.03**	**36.99**	**62.53**

In Table 7.3, the experimental results are provided on MELD, EmoryNLP, and IEMOCAP for different components in DBL. From Table 7.3, it is seen that BERT+CNN+Bi-LSTM can reach better performance on MELD, EmoryNLP, and IEMOCAP compared with BERT. As to these three data sets, the improvements of W-$F1$ are 2.32, 0.56, and 7.44%, respectively. The possible reason is that CNN and Bi-LSTM can help BERT+CNN+Bi-LSTM to extract the sequential information and long-distance dependencies.

From Table 7.3, it is observed that CBBL (i.e., BERT+CNN+Bi-LSTM+BL) can also reach better performance on MELD, EmoryNLP, and IEMOCAP compared with BERT. As to these three data sets, the improvements of W-$F1$ are 3.59, 2.07, and 8.38%, respectively. The possible reason is that CBBL can characterize the contextual relations of conversation by obtaining the utterance-level and speaker-level features.

In addition, CBBL can reach better performance on MELD, EmoryNLP, and IEMOCAP by comparing with BERT+CNN+Bi-LSTM. As to these three data sets, the improvements of W-$F1$ are 1.27, 1.51, and 0.94%, respectively. The possible reason is that BL can help CBBL to combine the utterance-level and speaker-level features effectively and then transform them into high dimension features.

7.3.6 *Ablation Analysis*

To comprehensively demonstrate the impact of utterance- and speaker-level encoding on CBBL, we conduct ablation analysis on MELD, EmoryNLP, and IEMOCAP by removing them separately, and analyze their contributions on these data sets. The experimental results are shown in Table 7.4.

From Table 7.4, it is seen that both utterance- and speaker-level encodings are crucial to the strong performance of CBBL on these data sets. There is a greater impact on ECTC task from the removal of utterance-level encoding than that of

Table 7.4 Ablation analysis with W-$F1$ for CBBL on MELD, EmoryNLP, and IEMOCAP

Utterance-level	Speaker-level	MELD	EmoryNLP	IEMOCAP
✓	✓	**64.03**	**36.99**	**62.53**
✓	×	59.84	35.32	62.18
×	✓	59.12	34.75	61.40

speaker-level encoding, which indicates that the contextual information between adjacent utterances is more essential for TEC task.

7.3.7 *Error Analysis*

Although CBBL can achieve a better performance, it is still hard to classify certain emotions on these three data sets. We provide two examples for the error results in Tables 7.5 and 7.6 from EmoryNLP and MELD, respectively.

From Tables 7.5 and 7.6, it is found that CBBL is difficult to distinguish two pairs of emotional types (i.e., "disgust" vs "fear", and "peaceful" vs "powerful"). There are two possible reasons for the model error. One possible reason is that these two pairs of emotions are very similar. The other possible reason is that there is only a small number of data available for these emotional texts. Thus, how to precisely distinguish very similar emotions and how to effectively classify emotions under the limited data are still two challenging works for TEC task.

Table 7.5 The error examples of CBBL from EmoryNLP

Conversation	True/Prediction
S1: Hello?	Peaceful/Neutral
S1: Ohh great!!!	Joyful/Joyful
S1: But, he loves Emily?	Joyful/Joyful
S1: Okay.	Powerful/Neutral
S2: All right, so, okay…	Peaceful/Neutral
S1: Right!	Powerful/Powerful
S2: So tell me about this girl?	Joyful/Neutral

Table 7.6 The error examples of CBBL from MELD

Conversation	True/Prediction
S1: I can't believe you would actually say that. I would much rather be Mr.Peanut than Mr.Salty.	Surprise/Neutral
S2: No way! Mr.Salty is a sailor, all right, he's got to be, like, the toughest snack there is.	Anger/Anger
S3: I don't know, you don't wanna mess with corn nuts. They're craaazy.	Neutral/Neutral
S4: Oh my God. You guys! You gotta come see this! There's some creep out there with a telescope!	Fear/Joy
S3: I can't believe it! He's looking right at us!	Surprise/Surprise
S5: Oh, that is so sick.	Disgust/Neutral
S1: I feel violated. And not in a good way.	Disgust/Anger

7.4 Summary

In this chapter, we verified the motivation of the combination of DL and BL for ECTC, and demonstrated that the combination of DL (i.e., CNN and Bi-LSTM) and BL could also be beneficially adopted in ECTC task, by comparing with the baseline methods, which usually capture utterance- and speaker-level contextual dependencies on the basis of the DL models. Specifically, we applied BERT, CNN, Bi-LSTM, and BL to capture utterance-level and speaker-level context dependencies simultaneously in textual conversations. With these two different representations, the local contextual information (i.e., utterance-level) and global contextual information (i.e., speaker-level) were well captured at same time. The experimental results have shown that our proposed method outperforms the baseline methods on most of the testing data sets in W-$F1$.

References

1. Busso, C., Bulut, M., Lee, C.C., Kazemzadeh, A., Mower, E., Kim, S., Chang, J.N., Lee, S., Narayanan, S.S.: Iemocap: Interactive emotional dyadic motion capture database. Language Resources and Evaluation **42**(4), 335 (2008)
2. Cui, Y., Che, W., Liu, T., Qin, B., Yang, Z.: Pre-training with whole word masking for chinese bert. IEEE/ACM Transactions on Audio, Speech, and Language Processing pp. 1–8 (2021)
3. Feng, S., Wei, J., Wang, D., Yang, X., Yang, Z., Zhang, Y., Yu, G.: Sinn: A speaker influence aware neural network model for emotion detection in conversations. World Wide Web **24**, 2019–2048 (2021)
4. Gao, Q., Cao, B., Guan, X., Gu, T., Bao, X., Wu, J., Liu, B., Cao, J.: Emotion recognition in conversations with emotion shift detection based on multi-task learning. Knowledge-Based Systems **248**, 108861 (2022)
5. Ghosal, D., Majumder, N., Gelbukh, A., Mihalcea, R., Poria, S.: COSMIC: Commonsense knowledge for emotion identification in conversations. In: Proceedings of the 2020 Conference on Empirical Methods in Natural Language Processing. pp. 2470–2481 (2020)
6. Hu, D., Wei, L., Huai, X.: Dialoguecrn: Contextual reasoning networks for emotion recognition in conversations. In: Proceedings of the 59th Annual Meeting of the Association for Computational Linguistics and the 11th International Joint Conference on Natural Language Processing. pp. 7042–7052 (2021)
7. Jiao, W., Yang, H., King, I., Lyu, M.R.: HiGRU: Hierarchical gated recurrent units for utterance-level emotion recognition. In: Proceedings of the 2019 Conference of the North American Chapter of the Association for Computational Linguistics. pp. 397–406 (2019)
8. Kang, L., Liu, J., Liu, L., Zhou, Z., Ye, D.: Semi-supervised emotion recognition in textual conversation via a context-augmented auxiliary training task. Information Processing and Management **58**, 102717 (2021)
9. Kim, Y.: Convolutional neural networks for sentence classification. In: Proceedings of the 2014 Conference on Empirical Methods in Natural Language Processing. pp. 1746–1751. Doha, Qatar (2014)
10. Li, J., Ji, D., Li, F., Zhang, M., Liu, Y.: HiTrans: A transformer-based context- and speaker-sensitive model for emotion detection in conversations. In: Proceedings of the 28th International Conference on Computational Linguistics. pp. 4190–4200 (2020)
11. Li, Q., Wu, C., Zheng, K., Wang, Z.: Hierarchical transformer network for utterance-level emotion recognition. Applied Science **10**(13), 4447 (2020)

12. Loshchilov, I., Hutter, F.: Fixing weight decay regularization in adam. In: Proceedings of International Conference on Learning Representations (2018)
13. Lu, X., Zhao, Y., Wu, Y., Tian, Y., Chen, H., Qin, B.: An iterative emotion interaction network for emotion recognition in conversations. In: Proceedings of the 28th International Conference on Computational Linguistics. pp. 4078–4088 (2020)
14. Ma, H., Wang, J., Lin, H., Pan, X., Zhang, Y., Yang, Z.: A multi-view network for real-time emotion recognition in conversations. Knowledge-Based Systems **236**, 107751 (2022)
15. Poria, S., Hazarika, D., Majumder, N., Naik, G., Cambria, E., Mihalcea, R.: Meld: A multimodal multi-party dataset for emotion recognition in conversations. In: Proceedings of the 57th Conference of the Association for Computational Linguistics. pp. 527–536 (2019)
16. Shen, W., Chen, J., Quan, X., Xie, Z.: Dialogxl: All-in-one xlnet for multi-party conversation emotion recognition. In: Proceedings of the Thirty-Fifth AAAI Conference on Artificial Intelligence. pp. 13789–13797 (2021)
17. Shen, W., Wu, S., Yang, Y., Quan, X.: Directed acyclic graph network for conversational emotion recognition. In: Proceedings of the 59th Annual Meeting of the Association for Computational Linguistics and the 11th International Joint Conference on Natural Language Processing. pp. 1551–1560 (2021)
18. Yang, Z., Dai, Z., Yang, Y., Carbonell, J., Salakhutdinov, R.R., Le, Q.V.: Xlnet: Generalized autoregressive pretraining for language understanding. In: Advances in Neural Information Processing Systems. pp. 5753–5763 (2019)
19. Zahiri, S.M., Choi, J.D.: Emotion detection on tv show transcripts with sequence-based convolutional neural networks. In: Proceedings of the Workshops of the Thirty-Second AAAI Conference on Artificial Intelligence. pp. 44–52 (2018)
20. Zhou, L., Gao, J., Li, D., Shum, H.Y.: The design and implementation of xiaoice, an empathetic social chatbot. Computational Linguistics **46**(1), 53–93 (03 2020)
21. Zhu, L., Pergola, G., Gui, L., Zhou, D., He, Y.: Topic-driven and knowledge-aware transformer for dialogue emotion detection. In: Proceedings of the 59th Annual Meeting of the Association for Computational Linguistics and the 11th International Joint Conference on Natural Language Processing. pp. 1571–1582 (2021)

Chapter 8
Rational Graph Attention Network and Broad Learning for Emotion Classification in Textual Conversations

Abstract Although CBBL was proposed in Chap. 7 and has achieved good results, there is still room for improvement of its performance. Thus, in this chapter, we propose a new method for ECTE task, named RGATBL, which is based on RGAT and BL. RGATBL aims to provide a better solution to capture contextual features and long-distance dependency by using RGAT, RoBERTa, and BL. It is also the first effort to combine RGAT and BL to classify emotion in textual conversations.

8.1 Introduction

ECTC focuses on identifying the emotion of each utterance in a textual conversation [9], which has become a popular research topic in recent years. With the popularization of GCN, more and more researchers begin to use it for solving ECTC task. As illustrated in Fig. 8.1, there are multiple speakers and utterances in a conversation, which can be represented as a directed graph. Thus, it is feasible to model and classify the emotion of each utterance from these speakers on the basis of graph theory.

Ghosal et al. [1] proposed a dialogue graph convolutional network (DialogueGCN) for the ECTC task. They employed Bi-GRU to sequential context, and adopted a GCN to capture the self- and inter-speaker dependencies among utterances. However, it does not take sequential information into account.

Zhong et al. [14] proposed a knowledge-enriched transformer (KET) to achieve better performance on multiple textual conversation data set. They applied hierarchical self-attention to encode contextual utterances and utilized a context-aware graph attention mechanism to incorporate commonsense knowledge. However, they fail to consider various relations in external knowledge base, which may cause the loss of semantic information.

Zhang et al. [13] proposed a knowledge aware incremental transformer with multi-task learning (KAITML) for the ECTC task. Firstly, they designed a dual-level graph attention mechanism to leverage commonsense knowledge, which augments the semantic information of the utterance. Then, they utilized incremental trans-

S. Peng, L. Cao, *Textual Emotion Classification Using Deep Broad Learning*, Socio-Affective Computing 11, https://doi.org/10.1007/978-3-031-67718-2_8

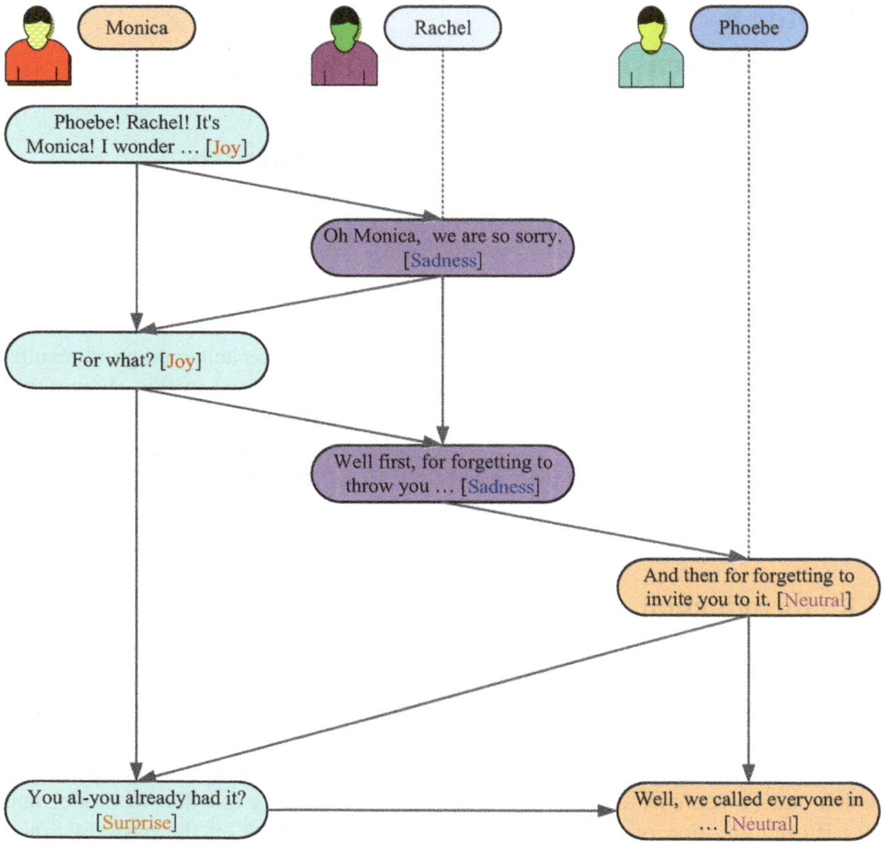

Fig. 8.1 A conversation as a directed graph

former to encode multi-turn contextual utterances. Finally, it introduced multi-task learning to alleviate the aforementioned confusion for improving its performance.

Ishiwatari et al. [5] proposed a RGAT-based framework with relational position encodings for the ECTC task. Firstly, they adopted BERT to obtain a contextual embedding for each utterance. Secondly, they utilized RGAT to modify this embedding for modelling speaker dependency, and considered the sequential information of the position encodings in RGAT structure. Finally, after concatenating contextual embedding to output embedding through RGAT, they applied a fully connected feed-forward network to classify the concatenated vector into emotion labels.

Heaton and Schwartz [3] integrated pre-trained language models (BERT and XLNet in their experiments) with a GCN. Firstly, it needs to identify the predicate-argument structure within an individual utterance. Then, it needs to construct for representing the predicate-argument structure identified in each utterance. As to each predicate-argument set in an utterance, a node is created for both the predicate and argument and an edge is added to connect the two nodes.

Shen et al. [10] designed a directed acyclic neural network, namely DAG-ERC, which adopts a directed acyclic graph (DAG) to encode the utterances, so as to better model the intrinsic structure in a conversation. DAG-ERC can combine the strengths of conventional graph-based neural models and recurrence-based neural models, and model the information flow between long-distance conversation background and nearby context.

In addition, some researchers adopted GCN to explore multi-modal emotion classification [2, 6–8] in conversation. Apart from GRU [2, 8], Bi-LSTM [6], and quantum-inspired neural network [7], there are some existing methods that explore to use GCN for modelling multi-modal emotion classification.

Shou et al. [12] proposed a model, called DSAGCN, by combining dependent syntactic analysis and GCN. Firstly, it employs a self-attention mechanism to capture the most effective words in the dialogue context and obtain a more accurate vector representation of the emotional semantics. Then, based on speaker relationships and dependent syntactic relationships, the multi-modal sentiment relationship graphs are constructed. Finally, a GCN is utilized to conduct the conversational emotion recognition.

Hu et al. [4] proposed an approach by using multi-modal fused GCN, called MMGCN. It can not only make use of multi-modal dependencies, but also model inter-speaker and intra-speaker dependency by leveraging speaker information. In MMGCN, the fully connected graph in each modality is constructed, and edge connections are built between nodes corresponding to the same utterance across different modalities, so as to interact contextual information across different modalities. In addition, the speaker information is injected into MMGCN via speaker embedding.

Zhang et al. [11] adopted a GCN to model both context-sensitive and speaker-sensitive dependence for ECTC task. At first, each utterance is denoted as a node in a graph, and an edge between two utterances in the same conversation to characterize the contextual dependence. Then, each speaker is represented as a node, and the specific-speaker dependence between each utterance and its speaker can be represented by an undirected edge. At last, the entire conversational corpus can be represented as a large heterogeneous graph and ECTC task can be formulated as a classification problem of the utterance nodes in the graph.

In the above mentioned methods, they usually adopt GCN to capture contextual information among utterances. To effectively capture contextual features and long-distance dependency, we utilize RGAT and BL for ECTC task in this chapter.

8.2 Methodology

RGATBL is a novel method for ECTC task. It consists of four components: utterance encoding based on RoBERTa, utterance-level context encoding based on Bi-LSTM, speaker-level encoding based on RGAT, and emotion classifier based on BL. The framework of DBL is shown in Fig. 8.2.

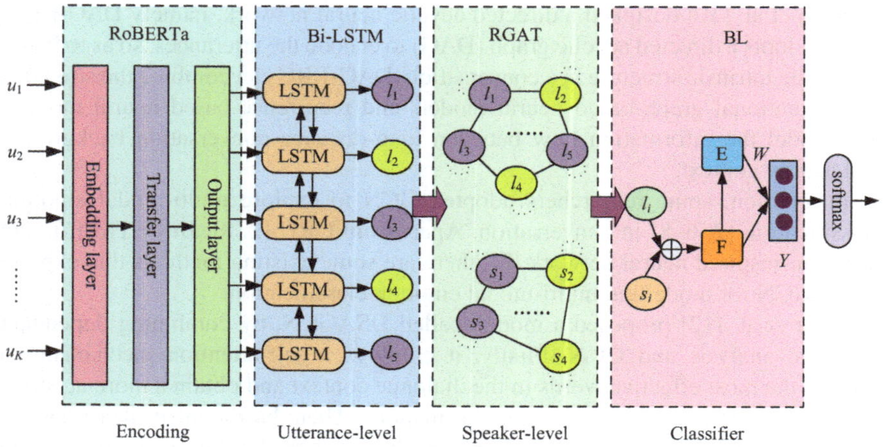

Fig. 8.2 The framework of the proposed method RGATBL, E denotes enhancement nodes, F denotes feature nodes, $u_1, u_2, \ldots, u_N$ denote utterances in a conversation

In the utterance encoding, RoBERTa pre-training model is adopted to obtain the word embedding of each utterance in a conversation. In the utterance-level context encoding and speaker-level encoding, Bi-LSTM and RGAT are exploited, respectively, to obtain two different types of sequence information. Finally, these features are input into the emotion classifier by using BL to obtain the prediction results.

8.2.1 *Text Encoding*

We use RoBERTa to extract the features and to generate word vectors for textual conversations; Then, we fine-tune RoBERTa to ensure that the parameters of each layer can satisfy the following relationships:

$$q_v^i = q_{v-1}^i - \mu^u \times \nabla_{q^l} H(q), \tag{8.1}$$

where q^i denotes the parameter of the i layer, v denotes the time step, $\nabla_{q^l} H(q)$ denotes the gradient of the model's objective function, and μ^u denotes the learning rate of u layer, which needs to satisfy the following relationship:

$$\mu^{u-1} = \delta \times \mu^u, \tag{8.2}$$

where δ denotes the decay rate of learning and $\delta \leq 1$.

When $\delta \leq 1$, the learning rate decays layer by layer, and when $\delta = 1$, the learning rate of each layer remains unchanged.

At last, we utilize the fine-tuned RoBERTa to pre-train each statement for generating a word vector, and take the output vector corresponding to the word symbol "[CLS]" as the sentence vector, so as to obtain the following sentence vector matrix:

$$U = [u_1, u_2, \ldots u_N] \in \mathbb{R}^{N \times d}, \tag{8.3}$$

where N denotes the number of utterances in a conversation, $u_i \in \mathbb{R}^d$ denotes sentence vector of i sentence, and d denotes the dimension of the sentence vector.

8.2.2 Utterance-Level Encoding

The advantage of Bi-LSTM is that it can effectively capture contextual features for extracting utterance-level features. The basic idea of Bi-LSTM is that it uses two LSTMs with the opposite directions to process forward and reverse sequences, respectively. The specific process is listed as follows:

$$\overrightarrow{l_i} = \overrightarrow{\mathrm{LSTM}}(\overrightarrow{l_{i-1}}, u_i), \tag{8.4}$$

$$\overleftarrow{l_i} = \overleftarrow{\mathrm{LSTM}}(\overleftarrow{l_{i+1}}, u_i), \tag{8.5}$$

$$l_i = [\overrightarrow{l_i}, \overleftarrow{l_i}], i = 1, 2, \ldots, N, \tag{8.6}$$

where $\overrightarrow{l_i}$ and $\overleftarrow{l_i}$ denote the forward and backward features of i sentence in LSTM, respectively.

Thus, as to a conversation, the utterance-level feature can be represented as follows:

$$L = [l_1, l_2, \ldots, l_N] \in \mathbb{R}^{N \times d_l}, \tag{8.7}$$

where, d_l denotes the dimension of utterance-level feature, and d_l denotes double number of neural units in a LSTM layer.

8.2.3 Speaker-Level Encoding

Because RGAT can effectively capture the structural features of textual conversation and the long-distance dependencies between two words, we adopt RGAT to conduct the speaker-level encoding. Firstly, we construct a directed graph to store the conversations and emotional interactions between two speakers. Then, we use RGAT to obtain the structural features of the textual conversations and the long-distance dependencies between two words. The specific construction process is described as follows.

We define a directed graph $G = \{V, E, R, W_G\}$ with K textual conversations, where V denotes node set, E denotes edge set, R denotes a set for the relationship type of edges, and W_G denotes a set for the attention weights of edges.

V: Each utterance $l_i \in V, i = 1, 2, \ldots, N$ matches a node of G, and each node is represented by the sentence vector $u_i \in \mathbb{R}^d$ generated by RoBERTa. Suppose $p(\bullet)$ denotes the mapping relationship between utterance and speaker, that is, $p(l_i) \in S$ denotes node l_i corresponding speaker $s_j, j = 1, 2, \ldots, M$.

E: If the edge between l_i and l_j is denoted by $e_{ij} \in E, i = 1, 2, \ldots, N, j = 1, 2, \ldots, M$, e_{ij} is utilized to denote the contextual relationship between two utterances. If it is needed to denote the contextual relationship between l_i and other utterance, l_i needs to connect all the nodes in G, that is, G becomes a fully connected graph.

R: There may be different contextual relationships between two utterances in different edges, including five types of relationships, which is denoted by $R = \{1, 2, 3, 4, 5\}$. When $p(l_i) = p(l_j), i < j$, $R = 1$; when $p(l_i) = p(l_j), i > j$, $R = 2$; when $p(l_i) \neq p(l_j), i < j$, $R = 3$; when $p(l_i) \neq p(l_j), i > j$, $R = 4$; when $i = j$, $R = 5$. Each edge represents the influence from other speakers or their own influence on the utterance of each speaker, so as to effectively extract the structural features of the textual conversations and the long-distance dependencies between two words.

W_G: The wight of edge e_{ij} between l_i and l_j can be denoted by $w_{ij} \in W_G$, and its value represents the impact degree from l_i to l_j.

The extraction process of speaker-level features based on RGAT is provided as follows.

We solve the attention weight w_{ij}, and adopt a single-layer feed-forward neural network to solve c_{ij} from l_j to l_i. The specific description is listed as follows:

$$c_{ij} = {l_i}^{\mathrm{T}} W_e \left[l_i, l_j\right], j = i - p, \ldots, i + f, \tag{8.8}$$

where W_e denotes the weight matrix.

To effectively represent the impact degree from different node to l_i, we adopt softmax to normalize the attention coefficient, and w_{ij} from l_j to l_i is represented as follows:

$$w_{ij} = \text{softmax}(c_{ij}) = \frac{\exp\left({l_i}^{\mathrm{T}} W_e \left[l_i, l_j\right]\right)}{\sum\limits_{k=i-p}^{i+f} \exp\left({l_i}^{\mathrm{T}} W_e \left[l_i, l_k\right]\right)}, j = i - p, \ldots, i + f. \tag{8.9}$$

We use two layers of RGAT to encode the speaker. In the first layer, l_i is transformed into the relevant feature vector h_i of a speaker by aggregating the information of neighbor nodes, and the specific process is represented as follows:

$$h_i = \sigma\left(\sum_{r \in \mathbb{R}} \sum_{j \in N_i^r} \frac{w_{ij}}{c_{i,r}} W_r^1 l_j + w_{ii} W_0^1 l_i\right), i = 1, 2, \ldots, N, \tag{8.10}$$

Table 8.1 The relationship among utterances from each speaker

Type	e_{ij}	Speaker	Relationship between i and j
5	e_{11}, e_{33}, e_{44}	s_1, s_1	$i = j$
1	e_{13}, e_{34}	s_1, s_1	$i < j$
2	e_{31}, e_{43}	s_1, s_1	$i > j$
5	e_{22}, e_{55}, e_{66}	s_2, s_2	$i = j$
2	e_{56}	s_2, s_2	$i < j$
1	e_{65}	s_2, s_2	$i > j$
3	$e_{12}, e_{35}, e_{45}, e_{46}$	s_1, s_2	$i < j$
4	e_{32}, e_{42}	s_1, s_2	$i > j$
3	e_{23}, e_{24}	s_2, s_1	$i < j$
4	$e_{21}, e_{53}, e_{54}, e_{64}$	s_2, s_1	$i > j$

where σ denotes the nonline activate function, W_r^1 and W_0^1 denote the weight matrix, N_i^r denotes l_i a set for the neighbor nodes under $r \in \mathbb{R}$, $c_{i,r}$ denotes normalization constant, usually sets $\left|N_i^r\right|$, which equals to the number of neighbor nodes of l_i.

In the second layer, the same feature transformation method is adopted on the basis of the first layer, that is, h_i is transformed to o_i. The specific process is represented as follows:

$$o_i = \sigma(\sum_{j \in N_i^r} W^2 h_j + W_0^2 h_i), i = 1, 2, \ldots, N, \tag{8.11}$$

where W^2 and W_0^2 denotes the weight matrices, p denotes the window size of historical utterance, and f denotes the window size of future utterance.

According to Eqs. (8.9) and (8.10), the information of neighbor nodes can get together, so as to obtain the long-distance dependency relationship between any two speakers, that is, speaker-level features can be represented as follows:

$$O = [o_1, o_2, \ldots, o_N] \in \mathbb{R}^{N \times d_s}, \tag{8.12}$$

where d_s denotes the feature dimension of speaker-level, that is, the number of hide units of RGAT.

For example, suppose there are six utterances $l_1, l_2, l_3, l_4, l_5, l_6$ in a conversation, where l_1, l_3, l_4 is generated by s_1, that is $s_1 = p(l_1) = p(l_3) = p(l_4)$; l_2, l_5, l_6 is generated by s_2, that is $s_2 = p(l_2) = p(l_5) = p(l_6)$. Suppose $p = 2$, $f = 2$, the relationship among utterances from each speaker is provided in Table 8.1.

8.2.4 Emotion Classifier

Firstly, the features of utterance- and speaker-level are concatenated, and based on the Eqs. (8.6) and (8.11), which can be obtained as follows:

$$G^* = [L|O] \in \mathbb{R}^{N\times(d_l+d_s)}. \tag{8.13}$$

Then, we adopt BL to design a emotion classifier for predicting emotional polarity of each utterance. The design process of emotion classifier is that G^* is linearly mapped into k groups of feature nodes, and the i-th group of feature nodes is represented as follows:

$$Z_i = \varphi(G^* W_{ei} + \beta_{ei}) \in \mathbb{R}^{N\times q}, i = 1, \ldots, k, \tag{8.14}$$

where φ denotes the linear activation function, W_{ei} denotes the weight matrix generated randomly and $W_{ei} \in \mathbb{R}^{(d_l+d_s)\times q}$, β_{ei} denotes the bias matrix generated randomly and $\beta_{ei} \in \mathbb{R}^{N\times q}$, q denotes the number of feature nodes in each group, and k denotes the number of groups of feature node.

Thus, k groups of feature nodes can be represented as follows:

$$Z^k = [Z_1, Z_2, \ldots Z_k] \in \mathbb{R}^{N\times kq}. \tag{8.15}$$

Then, Z^k is nonlinearly mapped into m groups of enhancement nodes, and the j-th group of enhancement nodes is represented as follows:

$$H_j = \xi(Z^k W_{hj} + \beta_{hj}) \in \mathbb{R}^{N\times r}, j = 1, \ldots, m, \tag{8.16}$$

where ξ denotes the nonlinear activation function, W_{hj} denotes the weight matrix generated randomly and $W_{hj} \in \mathbb{R}^{kq\times r}$, $\beta_{hj} \in \mathbb{R}^{N\times r}$ denotes the bias matrix generated randomly, and kq the number of enhancement nodes in each group, and r denotes the number of groups of enhancement node.

Thus, m groups of enhancement nodes can be represented as follows:

$$H^m = [H_1, H_2, \ldots, H_m] \in \mathbb{R}^{N\times mr}. \tag{8.17}$$

The matrix concatenated by Z^k and H^m is utilized to generate the actual input, which can be represented as follows:

$$A = [Z^k|H^m] \in \mathbb{R}^{N\times(kq+mr)}. \tag{8.18}$$

According to $\hat{Y} = AW$ and the matrix analysis theory, it can be obtained as follows:

$$W = A^+ Y. \tag{8.19}$$

Thus, to effectively solve W, the ridge regression is adopted as objective function which can be represented as follows:

$$\underset{W}{\operatorname{argmin}} \left(\left\| \left[Y - \hat{Y} \right] \right\|_2^2 + \lambda \left\| W \right\|_2^2 \right), \tag{8.20}$$

where λ denotes regularization parameters, and $\hat{Y} \in \mathbb{R}^{N \times (kq+mr)}$ denotes the final estimated output of BL.

W can be represented as follows:

$$W = (\lambda I + A^{\mathrm{T}} A)^{-1} A^{\mathrm{T}} Y. \tag{8.21}$$

8.3 Experiment and Analysis

In this section, we will provide a brief description for the data set, baseline methods, parameter settings, and evaluation metrics utilized in our experiments, and compare our proposed method with baseline methods. The experimental device is a laptop with Intel-i5-9400 2.9 GHz CPU and 16 GB memory, and server with Nvidia V100 GPU.

8.3.1 Data Set and Baseline Method

To verify the effectiveness of our proposed RGATBL, we conduct comparative experiments on data set like MELD, EmoryNLP, and IEMOCAP. Detailed information for each kind of data set has been introduced in Chap. 7. We compare the performance of our proposed method with the following baseline methods:

- **DialogueGCN** [1]: It is a model which uses GCN to capture the inter-speaker dynamics in a conversation. It also leverages the self and inter-speaker dependencies of the participants to model conversations.
- **KET** [14]: It is a transformer-based model which adopts external commonsense knowledge to improve contextual utterance representations.
- **KAITML** [13]: It is a graph-based model which introduces a dual-level graph attention mechanism and multi-task learning for conversational emotion recognition.
- **RGAT** [5]: It is a model which can capture both the speaker dependency and the sequential information by using RGAT.
- **DAG-ERC** [10]: It is an approach that utilizes RoBERTa and directed acyclic graph to solve ECTC task.

Table 8.2 W-$F1$ for different methods on MELD, EmoryNLP, and IEMOCAP

Methods	MELD	EmoryNLP	IEMOCAP
KET	58.18	34.39	59.56
KAITML	58.97	35.59	61.43
DialogueGCN	58.10	–	64.18
RGAT	60.91	34.42	65.22
DAG-ERC	63.65	**39.02**	68.03
DBL(Ours)	**65.93**	37.15	**69.02**

- **RGATBL**: It is our proposed method, which is a combination method based on RGAT and BL for ECTC considering utterance-level and speaker-level contextual information of textual conversation.

8.3.2 Performance Comparison

We evaluate the overall performance by using W-$F1$, and adopt the results of DialogueGCN reported in Ref. [1], the results of KET reported in Ref. [14], the results of KAITML reported in Ref. [13], the results of RGAT reported in Ref. [5], and the results of DAG-ERC reported in Ref. [10].

In Table 8.2, the experimental results are shown on MELD. It is seen that DAG-ERC can achieve the best overall performance of 63.65% on W-$F1$ score among the baseline methods. By comparison, the performance of our proposed RGATBL outperforms DAG-ERC by 2.28%. RGATBL can improve the overall performance, compared with the baseline methods.

From Table 8.2, it is found that DAG-ERC has the best overall performance among all the baseline methods. In particular, DAG-ERC performs better than our proposed method on IEMOCAP, which may be attributed to the capturing useful word- and utterance-level dependencies, and intra- and inter-speaker dependencies, respectively.

In Table 8.2, the experimental results are also shown on IEMOCAP. It is known that DAG-ERC achieves the best overall performance of 68.03% on W-$F1$ among all the baseline methods. By comparison, the performance of our proposed RGATBL outperforms DAG-ERC by 0.99%, which can prove the effectiveness of RGATBL. It also shows that there is a better capability of RGATBL to classify emotion in textual conversations.

8.4 Summary

In this chapter, we validated the motivation of the combination of RGAT and BL for ECTC, and demonstrated that the combination of RGAT and BL could also be beneficially adopted in ECTC task, by comparing with the baseline methods, which usually capture contextual features and long-distance dependency on the basis of RGAT and Bi-LSTM. It is shown that RGATBL is feasible by using RoBERTa, Bi-LSTM, and BL to capture the contextual features and long-distance dependencies simultaneously in textual conversations. With these two different representations, the utterance- and speaker-level information were well captured at the same time. The experimental results have shown that our proposed method outperforms the baseline methods on most of the testing data sets in W-$F1$.

References

1. Ghosal, D., Majumder, N., Poria, S., Chhaya, N., Gelbukh, A.: DialogueGCN: A graph convolutional neural network for emotion recognition in conversation. In: Proceedings of the 2019 Conference on Empirical Methods in Natural Language Processing and the 9th International Joint Conference on Natural Language Processing. pp. 154–164 (2019)
2. Hazarika, D., Poria, S., Mihalcea, R., Cambria, E., Zimmermann, R.: ICON: Interactive conversational memory network for multimodal emotion detection. In: Proceedings of the 2018 Conference on Empirical Methods in Natural Language Processing. pp. 2594–2604 (2018)
3. Heaton, C.T., Schwartz., D.M.: Language models as emotional classifiers for textual conversation. In: Proceedings of the ACM Multimedia. pp. 2918–2926 (2020)
4. Hu, J., Liu, Y., Zhao, J., Jin, Q.: Mmgcn: Multimodal fusion via deep graph convolution network for emotion recognition in conversation. In: Proceedings of the 59th Annual Meeting of the Association for Computational Linguistics and the 11th International Joint Conference on Natural Language Processing. pp. 5666–5675 (2021)
5. Ishiwatari, T., Yasuda, Y., Miyazaki, T., Goto, J.: Relation-aware graph attention networks with relational position encodings for emotion recognition in conversations. In: Proceedings of the 2020 Conference on Empirical Methods in Natural Language Processing. pp. 7360–7370 (2020)
6. Li, D., Li, Y., Wang, S.: Interactive double states emotion cell model for textual dialogue emotion prediction. Knowledge-Based Systems **189**, 1–11 (2020)
7. Li, Q., Gkoumas, D., Sordoni, A., Nie, J.Y., Melucci, M.: Quantum-inspired neural network for conversational emotion recognition. In: Proceedings of the Thirty-Fifth AAAI Conference on Artificial Intelligence. pp. 13270–13278 (2021)
8. Majumder, N., Poria, S., Hazarika, D., Mihalcea, R., Gelbukh, A., Cambria, E.: Dialoguernn: An attentive rnn for emotion detection in conversations. In: Proceedings of the AAAI Conference on Artificial Intelligence. vol. 33, pp. 6818–6825 (2019)
9. Peng, S., Zeng, R., Liu, H., Cao, L., Wang, G., Xie, J.: Deep broad learning for emotion classification in textual conversations. Tsinghua Science and Technology **29**(2), 481–491 (2024)
10. Shen, W., Wu, S., Yang, Y., Quan, X.: Directed acyclic graph network for conversational emotion recognition. In: Proceedings of the 59th Annual Meeting of the Association for Computational Linguistics and the 11th International Joint Conference on Natural Language Processing. pp. 1551–1560 (2021)
11. Shou, Y., Meng, T., Ai, W., Yang, S., Li, K.: Conversational emotion recognition studies based on graph convolutional neural networks and a dependent syntactic analysis. Neurocomputing **501**, 629–639 (2022)
12. Zhang, D., Wu, L., Sun, C., Li, S., Zhu, Q., Zhou, G.: Modeling both context-and speaker-sensitive dependence for emotion detection in multi-speaker conversations. In: Proceedings of the Twenty-Eighth International Joint Conference on Artificial Intelligence. pp. 5415–5421 (2019)

13. Zhang, D., Chen, X., Xu, S., Xu, B.: Knowledge aware emotion recognition in textual conversations via multi-task incremental transformer. In: Proceedings of the 28th International Conference on Computational Linguistics. pp. 4429–4440 (2020)
14. Zhong, P., Wang, D., Miao, C.: Knowledge-enriched transformer for emotion detection in textual conversations. In: Proceedings of the 2019 Conference on Empirical Methods in Natural Language Processing and the 9th International Joint Conference on Natural Language Processing. pp. 165–176 (2019)

Chapter 9
Summary and Outlook

Abstract In this chapter, we will discuss the challenges (i.e., emotion description, data imbalance, language imbalance, domain relevance, understanding short texts, implicit emotion analysis, and DL model training) and future research trends (i.e., multi-language emotion classification, multi-modal emotion classification, cross-domain emotion classification, emotion classification based on social network analysis, emotion classification based on big data analysis, automatic recognition of negative emotions in text, evolution analysis negative emotion, emotion cause extraction, and optimizing large language model) in TEC. This trend has drawn an increasing number of researchers away from traditional machine learning to DL for their scientific research.

9.1 Challenges of Emotion Analysis

Due to the rapid development of social networks and DL technology, unprecedented challenges to emotion analysis have been posed. Though many researchers have proposed potential solutions for some of the discussed issues, there are still many other open issues requiring further exploration and deep study [1, 6]. In this section, we summarize the challenges of emotion classification, and point out the future trends in this field.

9.1.1 Emotion Description

At present, there is no unified definition for emotion and no unified standard to classify emotions effectively and scientifically, which may affect the performance of emotion feature extraction in texts. However, because of the three unique components of human emotion (physiological arousal, subjective experience, and external expression), different fields possess different understandings. For example, social psychology, developmental psychology, and neuroscience deem it impossible for researchers to hold the same understanding of emotion [1]. They may understand

S. Peng, L. Cao, *Textual Emotion Classification Using Deep Broad Learning*,
Socio-Affective Computing 11, https://doi.org/10.1007/978-3-031-67718-2_9

emotion from their own perspectives or based on their judgments. Thus, it is difficult to provide an unified emotion definition and to determine an unified standard to accurately and effectively characterize human emotions.

9.1.2 Data Imbalance

The TEC task has made great progress in NLP. However, most existing works assume there are as many positive samples as negative samples, while positive and negative samples are often distributed unevenly in practice. Emotion analysis is a more fine-grained classification on the basis of the sentiment analysis, and most existing works also assume the balanced sample sizes for each emotion category, which is not consistent with reality [27]. For example, in the data set of NLPCC 2014-task 1, there are 6254 short texts for "happy" and 411 short texts for "fear". Thus, when those methods suitable for the balanced classification are adopted to deal with the unbalanced data, the analytical results often fail to achieve their intended effects, directly affecting the performance of emotion classification.

9.1.3 Language Imbalance

Most existing methods for TEC task are aimed at English texts. Some recent methods have focused on Chinese texts, but these methods are based on emotion dictionary or semantic knowledge base that rely on external resources of a specific language [29]. It is difficult to transfer English-based emotion classification methods to other languages (e.g., Italian, Japanese, German, and French). In addition, there are lots of minority languages in our world, such as Persian, Thai, Vietnamese, Indonesian, Hindi, Malay, Burmese, Norwegian, Finnish, Croatian, Slovenian, Czech, Bulgarian, Estonian, Danish, and so on. Their training and test sets for emotion classification are relatively scarce, particularly for the more uncommon language resources, begetting a serious negative impact on the research and performance of non-English language emotion classification methods.

9.1.4 Domain Relevance

Since there are various of domains and the same word may deliver different emotions in different domains. For example, the word "long" may deliver many different emotions, "long" in the domain of catering service reviews expresses a negative emotional tendency in relation to long waiting time, while "long" in the domain of smartphone reviews expresses a positive emotional tendency in relation to long battery standby time [2]. In addition, "apple" is a type of fruit in the fruit families, while in the electronics families, it represents a type of mobile phone. Thus, to effectively classify emotion in texts, the domain relevance of words must be considered. Cross-domain emotion classification presents numerous pressing problems for resolution.

9.1.5 Understanding Short Texts

Social networks limit the length of their commentary, making short text very common with its sparseness, non-standard use of words, and massive data instead of traditional long text. At the same time, insufficient contextual semantic information, single-word polysemy, and multi-word synonymy make the topic information extraction difficult to perform accurately, affecting the final emotion classification performance [9, 21]. For example, "It is hard to finish the task in time." vs. "It is raining hard. Don't go out until it stops.", there are many senses for word "hard". If we want to understand the actual sense of "hard", it needs the help of context in texts. Thus, understanding short texts is a very challenging task in NLP.

9.1.6 Implicit Emotion Analysis

Implicit emotion expressions indicate emotion expressions that does not contain emotion polarity but still convey clear human-aware emotion polarity based on context [7, 18]. The data set applied in existing methods, each sample contains emotion clauses. In these methods, emotion clauses are leveraged, especially emotion clauses usually contain emotion words, which can help to extract emotion cause clauses. However, in some reviews, obvious emotion words may not be utilized to express emotion, which makes the traditional TEC methods ineffective [23]. For example, "Tom is full when he think of the takeout ordered last week." Among them, "Tom is full" is actually a negative comment, and the cause for this comment is "takeout". This expression lacks explicit emotion words, which makes the expression more obscure. Thus, it is a challenging work to how to effectively mine the relationship between implicit emotion and cause, and to how to extract emotion causes in implicit emotion context.

9.1.7 DL Model Training

DL model training needs to continuously adjust relevant parameters. However, during DL model training, model training speed presents the biggest problem due to slow convergence speeds and long training times. Thus, model training efficiency deserves consideration. Improving model convergence speed requires reduced iterations and consistent training times, while improving training speeds requires a reduced number of training passes and, consequently, reduced opportunity to attempt differing super parameters. Both improvements will affect model accuracy. In addition, training model performance relates to training data set size, with larger data sets producing better results [11]. However, larger scale data sets may increase the training time and require larger amounts of computer resources (e.g., GPU, memory, and CPU).

9.2 Future Research Trends

As Internet +, AI +, 5G, and other opportunities arise, many new applications (e.g., multi-language, multi-modal, cross-domain, big data, large model) have emerged and provided more opportunities for emotion classification. As emotion classification plays an important role in grasping public sentiment trends quickly, predicting public opinions of relevant development trends, and satisfying human daily requirements, emotion classification will change qualitatively by integrating different media, forms, scales, and domains of emotional information. In addition, the rapid development of social network analysis and DL technology offer many new research directions for emotion analysis. Some researchers are gradually changing the focus of research in emotion classification, from single language, single modal, single domain, small-scale data samples, and small model to multi-language, multi-modal, cross-domain, big data, large model [12]. According to existing technology development trends, future emotion classification research will include the following aspects.

9.2.1 Multi-Language Emotion Classification

Due to increasing cultural exchanges, numerous of multi-language mixed texts (e.g., English and Chinese, Japanese and Chinese) have been emerged in social networks [22]. However, most existing works have focused on a single language and the corpus resource collected for the single-language emotion classification model is difficult to satisfy the multi-language environment. In addition, these models cannot represent multi-language in the same feature space, which may lead to semantic bias. Thus, how to represent words with the same semantics in different languages to positions with similar spatial distances is the basis for solving multi-language TEC task.

9.2.2 Multi-Modal Emotion Classification

Traditional emotion classification focuses on single forms of media (e.g., text-based or audio-based or image-based), while multi-modal information (e.g., audio, video, and image) [8, 16] can often express emotional effect with greater description and vividness than text. In addition, as a main carrier of emotional information expression, audio and video can accurately reflect current user emotions. For example, emotion classification in conversation is a crucial component in emotion dialogue systems, which helps to understand users' emotions and generate empathetic responses [4, 5]. However, most existing methods focus on modeling speaker and contextual information primarily on the textual modality or simply leveraging multi-modal information through feature concatenation. Thus, exploring a more effective way to use both multi-modal and long-distance contextual information will become a new research task.

9.2.3 Cross-Domain Emotion Classification

The main idea of cross-domain emotion classification method is that emotions presented in current comment information can be identified accurately and quickly, provided that information contains words expressing various emotions from different domains [30]. However, traditional emotion classification methods often ignore domain dependency characteristics of emotion words, and may even deliberately choose domain-independent features (such as emoji). With increasing demand in practical applications and the emergence of emotion corpus resources in different domains, cross-domain emotion classification will attract greater research attention and focus.

9.2.4 Emotion Classification Based on Social Network Analysis

With the rapid development of social networks, a large amount of user review and interaction data has been generated. These data not only reflect static user characteristics (e.g., number of friends, activity, frequency of surfing the Internet), but also reflect dynamic user characteristics (e.g., thoughts, social relations, social influence) [13, 15]. Through the analysis of social networks, we can understand how different individuals and social groups express their emotions and how group emotional tendencies relate to the popular events. Therefore, research on emotion classification technology based on the social network analysis can better grasp public opinion trends while providing a technical support for applications involving big search, public opinion analysis, personalized recommendation, negative propagation, and so on.

9.2.5 Emotion Classification Based on Big Data Analysis

With the increasing scale of various of social networks, massive data are generated every day. Mining these data can provide very valuable information for our products and services. However, a significant portion of that data may be stored in an unstructured manner after being collected by crawlers. When emotion classification is conducted on these text data, traditional methods such as probabilistic latent semantic analysis may be difficult to satisfy the needs of these large-scale data training. This will push some new methods to be presented for emotion classification on the basis of big data analysis [10, 14].

9.2.6 Automatic Recognition of Negative Emotion in Short Texts

At present, there are a large number of comments on Wechat, Facebook, Twitter, Taobao, and other social networks [20]. Much of the massive number of short text contains negative emotion [3], which makes the automatic identification of negative emotion from such recognition an urgent requirement along with designing some more effective methods to extract the negative emotion features. Such requirements will play an important role for maintaining the national cyberspace security, and will attract more and more researchers to study automatic negative emotion recognition in short texts.

9.2.7 Negative Emotion Evolution Analysis

Due to expression manners, some popular events may lead to negative emotion with the highest concentration of these emotions being emerged in social networks. Such networks (e.g., Microblog, Twitter) usually have propagation characteristics of "weak information and strong emotion", which can result in negative emotions propagated widely across Microblog or Twitter. Once the information with negative emotions is released, it may propagate by means of nuclear fission. If it is amplified by users with great social influence (e.g., opinion leaders, hostile forces) [24], it will cause the public opinion or political event. Thus, characterizing the internal evolution of negative emotions for popular events will become a noteworthy issue of TEC and cyberspace security.

9.2.8 *Emotion Cause Extraction*

Emotion cause extraction (ECE) aims to automatically recognize the potential cause or stimulus of an observed emotion in texts [17, 26]. ECE is an important research issue in natural language processing and is different from the previous tasks of emotion recognition and emotion classification. In addition, it is not limited to the shallow-level emotion classification of text, but to trace the emotion source. Thus, ECE is an in-depth emotion analysis method that focuses on emotional keywords in text to identify causes automatically. Although current mainstream methods are based on linguistic rules and statistics, the wide application of DL, large language model, and BL will continue to attract increasing attention in ECE research.

9.2.9 *Optimizing Large Language Model*

Although ChatGPT [25] is one of the most successful large language model, it is difficult to design language algorithms and lack of machine thinking skills. In addition, BERT neglects dependency among predicted tokens. XLNet [28] does not leverage full position information in sentence, and then results in position discrepancy between pre-training and fine-tuning. MPNet [19] inherits the advantages of BERT and XLNet, and avoids their limitations. Although many optimized pre-training models have been proposed, it is difficult to train and is inefficient due to their numerous parameters. Thus, it is necessary to optimize the existing pre-training models.

9.3 Summary

In this chapter, we mainly share our understanding on emotion classification task in social networks in the near future for readers. As it is difficult to predict the picture of tomorrow, readers do not necessarily need to follow the suggestions but to refer to.

In the future, we may face lots of unprecedented problems and challenges, but aggressive researchers are hopeful for theoretical solutions. We humbly hope that the content of this chapter will help the interested readers to explore some relevant fields in the future.

References

1. Cao, L., Peng, S., Yin, P., Zhou, Y., Yang, A., Li, X.: A Survey of Emotion Analysis in Text Based on Deep Learning. In: Proceedings of the IEEE 8th International Conference on Smart City and Informatization (iSCI 2020). pp. 81–88. Guangzhou, China (2020)
2. Cao, Z., Zhou, Y., Yang, A., Peng, S.: Deep transfer learning mechanism for fine-grained cross-domain sentiment classification. Connection Science **33**(4), 911–928 (2021)
3. Chen, G., Peng, S., Zeng, R., Hu, Z., Cao, L., Zhou, Y., Ouyang, Z., Nie, X.: p-norm broad learning for negative emotion classification in social networks. Big Data Mining and Analytics **5**(3), 245–256 (2022)
4. Hu, J., Liu, Y., Zhao, J., Jin, Q.: Mmgcn: Multimodal fusion via deep graph convolution network for emotion recognition in conversation. In: Proceedings of the 59th Annual Meeting of the Association for Computational Linguistics and the 11th International Joint Conference on Natural Language Processing. pp. 5666–5675 (2021)
5. Li, Q., Gkoumas, D., Sordoni, A., Nie, J.Y., Melucci, M.: Quantum-inspired neural network for conversational emotion recognition. In: Proceedings of the Thirty-Fifth AAAI Conference on Artificial Intelligence. pp. 13270–13278 (2021)
6. Li, R., Lin, Z., Lin, H., Wang, W., Meng, D.: Text emotion analysis: a survey. Journal of Computer Research and Development **55**(1), 30–52 (2018)
7. Li, Z., Zou, Y., Zhang, C., Zhang, Q., Wei, Z.: Learning implicit sentiment in aspect-based sentiment analysis with supervised contrastive pre-training. In: Proceedings of the Conference on Empirical Methods in Natural Language Processing (EMNLP 2021). pp. 246–256 (2021)
8. Liu, B., Tang, S., Sun, X., Chen, Q., Cao, J., Luo, J., Zhao, S.: Context-aware social media user sentiment analysis. Tsinghua Science and Technology **25**(4), 528–541 (2020)
9. Lu, H., Yang, J., Zhang, Y., Li, Z.: Polysemy needs attention: Short-text topic discovery with global and multi-sense information. IEEE Access **9**, 14918–14932 (2021)
10. Mahmud, M., Huang, J., Salloum, S., Emara, T., Sadatdiynov, K.: A survey of data partitioning and sampling methods to support big data analysis. Big Data Mining and Analytics **3**(2), 85–101 (2020)
11. Peng, S., Cao, L., Zhou, Y., Xie, J., Yin, P., Mo, J.: Challenges and Trends of Android Malware Detection in the Era of Deep Learning. In: Proceedings of the IEEE 8th International Conference on Smart City and Informatization (iSCI 2020). pp. 37–43. Baltimore, MD, USA (2020)
12. Peng, S., Wang, G., Xie, D.: Social influence analysis in social networking big data: opportunities and challenges. IEEE Network **31**(1), 11–17 (2017)
13. Peng, S., Wang, G., Zhou, Y., Wan, C., Wang, C., Yu, S., Niu, J.: An immunization framework for social networks through big data based influence modeling. IEEE Transactions on Dependable and Secure Computing **16**(6), 984–995 (2019)
14. Peng, S., Yang, A., Cao, L., Yu, S., Xie, D.: Social influence modeling using information theory in mobile social networks. Information Sciences **379**, 147–159 (2017)
15. Peng, S., Zhou, Y., Cao, L., Yu, S., Niu, J., Jia, W.: Influence analysis in social networks: a survey. Journal of Network and Computer Applications **106**, 17–32 (2018)
16. Peng, W., Hong, X., Zhao, G.: Adaptive modality distillation for separable multimodal sentiment analysis. IEEE Intelligent Systems p. DOI: 10.1109/MIS.2021.3057757 (2021)
17. Qiu, X., Liu, D., Wan, C., Li, J., Liu, X., Liao, G.: Survey on automatic emotion cause extraction from texts. Journal of Computer Research and Development **59**(11), 2467–2496 (2022)
18. Russo, I., Caselli, T., Strapparava, C.: Semeval-2015 task 9: Clipeval implicit polarity of events. In: Proceedings of the 9th International Workshop on Semantic Evaluation (SemEval 2015). pp. 443–450. Denver, Colorado (2015)
19. Song, K., Tan, X., Qin, T., Lu, J., Liu, T.Y.: Mpnet: Masked and permuted pre-training for language understanding. In: Advances in Neural Information Processing Systems. vol. 33, pp. 16857–16867 (2020)

20. Song, X., Jiang, W., Liu, X., Lu, H., Tian, Z., Du, X.: A survey of game theory as applied to social networks. Tsinghua Science and Technology **25**(6), 734–742 (2020)
21. Sun, Y., Yang, Y., Yang, D.: Informed graph convolution networks for multilingual short text understanding. In: Proceedings of the 26th International Conference on Knowledge-Based and Intelligent Information & Engineering Systems (KES 2022). pp. 90–99 (2022)
22. Suzuki, M., Qi, J.: Improvement of multilingual emotion recognition method based on normalized acoustic features using CRNN. In: Proceedings of the 26th International Conference on Knowledge-Based and Intelligent Information & Engineering Systems (KES 2022). pp. 684–691 (202)
23. Wei, J., Liao, J., Yang, Z., Wang, S., Zhao, Q.: Bilstm with multi-polarity orthogonal attention for implicit sentiment analysis. Neurocomputing **383**, 165–173 (2020)
24. Wu, J., Wang, N.: Approximating special social influence maximization problems. Tsinghua Science and Technology **25**(6), 703–711 (2020)
25. Wu, T., He, S., Liu, J., Sun, S., Liu, K., Han, Q.L., Tang, Y.: A brief overview of chatgpt: The history, status quo and potential future development. IEEE/CAA Journal of Automatica Sinica **10**(5), 1122–1136 (2023)
26. Xia, R., Zhang, M., Ding, Z.: Rthn a rnn-transformer hierarchical network for emotion cause extraction. In: Proceedings of the Twenty-Eighth International Joint Conference on Artificial Intelligence (IJCAI 2019). pp. 5285–5291 (2019)
27. Xu, R., Chen, T., Xia, Y., Lu, Q., Liu, B., Wang, X.: Word embedding composition for data imbalances in sentiment and emotion classification. Cognitive Computation **7**, 226–240 (2015)
28. Yang, Z., Dai, Z., Yang, Y., Carbonell, J., Salakhutdinov, R.R., Le, Q.V.: Xlnet: Generalized autoregressive pretraining for language understanding. In: Advances in Neural Information Processing Systems. pp. 5753–5763 (2019)
29. Yilmaz, S.F., Kaynak, E.B., Koc, A., Dibeklioglu, H., Kozat, S.S.: Multi-label sentiment analysis on 100 languages with dynamic weighting for label imbalance. IEEE Transactions on Neural Networks and Learning Systems **34**(1), 331–343 (2023)
30. Zhao, C., Wang, S., Li, D.: Research progress on cross-domain text sentiment classification. Journal of Software **31**(6), 1723–1746 (2020)

GPSR Compliance

The European Union's (EU) General Product Safety Regulation (GPSR) is a set of rules that requires consumer products to be safe and our obligations to ensure this.

If you have any concerns about our products, you can contact us on ProductSafety@springernature.com

In case Publisher is established outside the EU, the EU authorized representative is:

Springer Nature Customer Service Center GmbH
Europaplatz 3
69115 Heidelberg, Germany

The manufacturer's authorised representative in the EU is Springer Nature Customer Service Centre GmbH, Europaplatz 3, 69115 Heidelberg, Germany. If you have any concerns regarding our products, please contact ProductSafety@springernature.com

Printed and bound by CPI Group (UK) Ltd, Croydon, CR0 4YY
01/10/2026
02225916-0016